THE POLITICS OF CATACLYSM

Scientists and politicians usually agree when it comes to concern about public reaction to danger or disaster. The secret is to keep the populace calm and quiet . . . camouflage the facts, confuse the problem with rhetoric and statistics. Witness the recent fiascos regarding the so-called Legionnaire's Disease and Swine Flu.

And what about natural catastrophes involving the Earth and its weather? Think about recent headline stories of earthquakes, floods, and hurricanes. With all our technology and expertise (and propaganda) we still cannot control or predict these deadly events. Thousands of lives and millions of dollars in damage become yesterday's news and tomorrow's footnotes.

Earthquakes are sudden. Tornadoes and typhoons are sudden. Cataclysms have been—and will be—sudden. There will be tremendous losses of life, but millions of people as well as animals and plants will survive, even though millions may perish. The very thought of cataclysmic disaster is so horrifying that it could well be one reason why traditional geologists would rather not even consider it possible. However, it is possible; it has happened.

Yes, there may even be extinctions of some species of animals. In fact, cataclysm is a far more logical reason for the great extinctions of species of the distant past, such as dinosaurs and mammoths, than the rationalizations drummed up by uniformitarians. That's where we go from here. We had better make sure our view of the past is correct and not merely construed to fit a comfortable paradigm.

Pinnacle Books by Tom Valentine:

THE GREAT PYRAMID
THE LIFE & DEATH OF PLANET EARTH

The LIFE & DEATH of PLANET EARTH

by Tom Valentine

PINNACLE BOOKS   LOS ANGELES

THE LIFE & DEATH OF PLANET EARTH

An original Pinnacle Books edition, published for the first
time anywhere.

ISBN: 0-523-40960-5

First printing, April 1977
Second printing, December 1977
Third printing, July 1980

Cover illustration by Larry Kresek

Printed in the United States of America

PINNACLE BOOKS, INC.
2029 Century Park East
Los Angeles, California 90067

The illustrations in the sixteen page insert are reproduced
by courtesy of the author.

This book is dedicated to an organization and a person. The organization is The Stelle Group, *a gathering of courageous people who are striving toward the ultimate frontier of human perfection and an optimum civilization. The person is Kathy Redmond, who worked tirelessly to curb my over-exuberance, to uplift my downers, to sharpen my syntax, and generally to bring about a better book.*

It is vital that each reader fully understand that, although I am a member of *The Stelle Group,* I do not necessarily speak for the other members or the governing body. The conjecture in this book is entirely my own and does not necessarily correspond with the thinking of my organization. It is a tribute to the spirit of our community that each individual member is not only permitted, but encouraged, to think for himself and express his individualism. In this organization, though we all seek the same goals of egoic advancement and a better civilization, we each maintain total individuality.

PROLOGUE

You, The Jury!

I am neither prophet nor religious fanatic. I'm a journalist, and I have facts to share with you much as a lawyer does when making a case before a panel. Essentially, I'm simply another human being trying to separate fact from fiction so I can determine a personal course of action for the future. As a journalist I've probably listened to more varied beliefs, opinions, ideas, notions, theories, and solutions than 99.99 percent of the populace. I've learned much from listening to others. I've learned that society floats in a sea of entrenched error; we are surrounded by accepted nonfact. Our schools teach error, our churches put forth assumptions based on error, and our solutions to social problems rest on fallacious notions.

The thesis of this book is that we are indeed living in the Biblical "times of the end." A violent cataclysm, which will rearrange the continental plates that make up the crust of the Earth, looms for the turn of the century. This thesis is based on hard scientific fact as well as religious prophecy. My plea, then, is for my jury of readers to remain open-minded and hear me out, lest our technically sophisticated civilization be suddenly submerged beneath mud, boulders, lava, and oceans. For if nothing is done to prepare for the impending disaster, there is a good chance that our times will vanish, only to become the "legendary" times for civilizations of the future to debate; much as we debate the existence of Atlantis today.

Despite the forecast of violent geophysical changes, I'm not a prophet of gloom. The thesis put forth here advances to become one of promise for mankind, providing mankind gets off its collective duff and strives to turn disaster into opportunity. The catastrophic consequences and the appalling death toll that would result from this natural disaster do not necessarily mean the the end of the world. Quite the contrary! Such a thorough housecleaning can mean the end to many unsolvable social problems and a fresh, positive start toward rebuilding. A new civilization may be ushered in—perhaps a civilization that fulfills the prophecy of a "nation of God."

This book makes a case for such conjecture. Here is a rational, logical, systematic approach to the age-old problem of planning realistically for the future in the midst of religious fervor, academic dogma, popular skepticism, and the pressures of daily living. This book is a serious attempt to begin a rewriting of erroneous views of history and geology, and to promote a sensible approach to what lies ahead. There isn't much time in which to wage a successful campaign. The date I envision for cataclysm is May 5, 2000—a scant twenty-three years hence.

As I see it, it's a matter of life and death. The life and death of Planet Earth and everything humanity has ever achieved. This thesis is presented to you much as one would present facts to a panel at a hearing or a grand jury. Upon conclusion of this hearing, a decision on a course of action must be rendered by you, the jury.

TABLE OF CONTENTS

The Life & Death of Planet Earth

CHAPTER 1

Why We Base Our Lives on Faulty Beliefs

Though it's only vague memory now, I do recall how we roared with laughter upon hearing the reason one family from our sleepy California hometown was leaving the Golden State to move to Colorado. I had just graduated from high school in the summer of 1953, and several of us were gathered at the ballpark one warm afternoon discussing the impending exodus of two of our schoolmates because their parents believed California was going to break off from the continent of North America and plunge into the sea. There were at least seven members of our American Legion baseball team taking part in the conversation, and each of us was absolutely certain that the family we were subjecting to arbitrary ridicule was somehow looney, while we, on the other hand, were sane and sensible.

Since California still hasn't dipped into the Pacific, at least not as of this writing, one might surmise that we schoolboys were right in our opinion that the departing family was kooky. However, on looking back, I now ask myself: "On what body of fact were we making our judgment? What made a group of teen-age boys so certain that nothing so catastrophic as California vanishing into the sea could happen?"

Of course, the subject was never really debated or discussed. We simply ridiculed that family because they had an unusual set of beliefs. People do, in fact, operate their daily lives on what they *believe*. After all, what do any of us really *know* about the beginnings of this solar system and this planet? What do we really *know* about the origins of life? What do we really *know* about prehistory and man's early stages? What do we really *know* about recorded history? We accept the historian's chronicle of events, even though abundant proof of bias has been uncovered and we realize that history has every chance of becoming biased each time it is rewritten or a new scholarly point of view is injected into the record.

In truth, we *know* very little, but we *believe* we know a great deal. Where did we get this overstuffed idea? From our institutions, our experts, our textbooks, that's where! As we mature and become more educated, we resort to the collective grab bag of expertise called "science" to cull our factual information. We have established a pattern for testing information—or hypotheses—called the "scientific method," which helps determine reliable facts. Invariably the method begins with a problem statement, progresses through the gathering of relevant data, and subsequently requires the formulation of an hypothesis to be tested. If the hypothesis proves out and conforms to the existing body of scientific knowledge, the result is a theory or conclusion to the testing—the problem has been solved, at least for the time being.

Somehow, however, the true meaning of that last step has become so obscured that we have allowed *theory* to become fact in a number of scientific areas. It is

2

all too convenient to forget that in step three a limitless number of hypotheses can be entertained for testing, that a conclusion born out by testing "temporary" hypotheses is always open to question and, therefore, our tendency to misconstrue a theory and let it be treated as fact is erroneous.

Admittedly, it is far easier to function from a framework of fact than of theory, so the penchant for passing theory off as dogma is a strong one. We live under a dogmatic umbrella of accepted beliefs about our planet and man's history which shapes much of our behavior, yet these beliefs are *wrong*!

For example, the majority of Americans *believe* that this orbiting, spinning lump of cosmic real estate is untold millions of years old; that it became a solid body by cooling down from hot gases; that the continents are stable and move only slightly over long periods of time; that somehow weather patterns have changed drastically in the past to allow huge sheets of ice to cover vast areas of temperate lands (ice ages); that God may have created it all, but He obviously used the evolutionary process over millions of years to bring about His favored creature—man; that man slowly developed thinking processes over at least a million years to reach a point a few thousand years ago where he learned agriculture and, thus, basic civilization; that mankind stepped from a stone age culture to the dawning of civilization in river valleys of the Middle East, India, and China a mere 7,000 years ago; and finally, it is believed that we, the twentieth century people, are the best brand, culturally and technically, that mankind has yet produced.

A few years ago if you challenged that body of information and had the guts to admit your challenge, you were branded a raving lunatic by your neighbors—as we teen-agers had done to our neighbor. Today, if you don't believe all of that accepted earth-man chronology, you are merely considered part of a kook

3

minority that hasn't anything better to do with its time than to take potshots at what "everyone knows" is truth.

One thing is certain: We do *know* that our world is far from perfect, and if anything is going to be done to improve the situation, I believe we must revamp our thinking. To do this it is necessary that we purge as much error and ignorance from our views of our heritage and this planet as we possibly can. From this moment on there are no infallible experts, regardless of the credentials that might have been collected from our institutions of higher learning. We are about to strip fact from fancy. This is not to say that everyone is stupid. This will merely allow us to remove presumed omnipotence from a body of thought and a body of thinkers to enable us to reevaluate all our concepts.

Generally speaking, our sciences are not exact disciplines. Geometry and mathematics come very close, but still many equations and concepts are approximate or theoretical. Most disciplines, especially those ending in the suffix "ology," are nothing more than picture-puzzle studies with many pieces that don't fit. Geology, archaeology, paleontology, anthropology, and the like all have dominant *theories* which have evolved into *accepted* fact, even though they are far from *proven* fact.

However, in order to function in our daily lives, which is what each of us generally wants to do, certain frameworks of understanding must be constructed around each scientific discipline, much like rules of a game. Technology, for example, requires certain hard and fast rules, mainly to stimulate advancement; if one had to begin from the very beginning to establish a framework of reference—or reestablish facts that are already known to science—advancement would take place only at the end of a lifetime if at all. This is simply too time-consuming for the pace of civilization. On the other hand, these hard and fast rules also account for engineers usually being able to explain why some-

4

thing won't work more readily than why something will work. Because our technology is so sophisticated, thanks to a mysterious and partially controllable energy called electricity, we tend to believe we know much more than we really do.

A few hundred years ago, Roger Bacon, the brilliant English scholar, stated it succinctly: "Man can *do* much more than he *knows*." (Emphasis added.) Microbiologists have never really seen a gene, but they can manipulate them. The experts still aren't certain about what electricity really is, but we all flip switches.

It is important for us to understand that when technologists want to draw a line around a scientific school of thought and give it discipline, they construct a framework that includes workable theory, experimentation, and practical experience. This framework is then called a "paradigm." For some reason—most likely the puffery associated with the human ego—paradigms tend to become absolute and dogmatic. Then, whenever some freethinker (often a genuine genius) comes along and pokes holes in the paradigm, a body of defenders who are usually cloaked in bureaucratic or priestly robes, will rise up and indignantly attempt to suppress the threatening new ideas. The most effective way to defend paradigms from genius, the defenders have learned, is to attack the personality and ignore the facts.

Sometimes the defenders of the faith, or dogma, are ludicrous in retrospect—like the "experts" telling the Wright brothers and others that "it will never fly." Other times the defense of dogma has become tragic and hateful, such as when the scions of the church put Bruno to the stake for pointing out that the Aristotelian paradigm regarding relative positions of the sun, earth, and other planets was incorrect.

Sometimes anomalies (a word used to describe those nagging little pieces of fact that don't fit the puzzle properly) force scientists to alter their paradigms with-

5

out the general public catching on for years. Such has been the case with geology, the study of the crust of the earth.

Most of our schools still teach our children the nebular theory to explain the earth's origins. This theory explains that the earth cooled from hot gases to become a hard little ball of matter. Although it is still taught in public schools, specialists know that the nebular theory was superseded by the planetesimal theory a number of years ago. The change in thinking came about because the latter supposition holds up more responsibly to the dictates of known chemistry and physics. It states that the earth may have originated as a small planetesimal and slowly grew by attracting particles, such as meteorites and meteorite dust. It may have grown hot because of the internal pressures caused by the increasing mass, and because of the various effects of radioactive bits of matter.

The profound effect that this radical change in theory should have had on science was expressed back in the twenties by William H. Hobbs, an eminent geologist, who said: "Far more than is generally supposed, the recent abandonment of the nebular hypothesis to account for the origin of the universe must carry with it a rewriting of our science. This is particularly true of geology, for all that concerns seismology, volcanology, and the whole subject of the growth of continents and mountains . . . ."

There may have been a profound effect somewhere, but there has never been a rewriting of the geological paradigm. Even today some geologists convey to the public the idea that the question of the earth's cooling has never been questioned. So you can plainly see that many scientific assumptions in geology are still imbedded in a worn-out, unworkable theory.

So far the only point being made is that geologists have changed their minds about how the earth was formed, and most of us don't think that's any big deal.

6

However, the science of geology deals with the very foundation of our lives—the crust of the earth upon which we live. Is that crust as solid and stable and permanent as it seems to be to us?

Most people would assume that, except for an occasional earthquake or volcano, the earth is stable and permanent, and like a bunch of teen-age ballplayers, they would laugh at the thought of pulling up stakes and leaving a home because a psychic or "kook" geologist predicted a cataclysm. On the other hand, what would you do if tomorrow your city, or town, or farm suddenly rumbled, rocked, heaved, and sloshed to another location on the globe. In light of the changing paradigm of geology, it seems we should stop laughing and consider the possibilities. As one friend of mine said, "I'll bet the Atlanteans laughed all the way to the bottom of the Atlantic."

Don't say it can't happen to us, because it most certainly can. It has happened before! If, however, you wish to be skeptical and count on the science of geology to support your skepticism, just remember that once you skim off the forsaken theories and bare only the facts, there is no support for your ridicule.

Let's analyze what we do *know* about the crust of our planet. We know that earthquakes and volcanoes occur. We know there are mountain ranges and deep canyons and places that have been scarred by icy glaciers.

We don't know what causes mountain building or ice ages, but the fact of such things tells us that our continents are anything but stable. One day in 1943, for example, there was a level cornfield in Mexico. Then, all of a sudden, there were a few rumbles, snaps, crackles, pops, and Paricutin erupted. A short while later we have Earth's newest mountain 8,700 feet high, a half-mile higher than anything in the Appalachians. Since we don't live in that part of Mexico, we thought this was an interesting news item and nothing more.

7

The face of earth is constantly changing, constantly shifting and erupting or sinking. And yet the vast majority of people will deny the possibility of cataclysmic change, just as my friends and I denied the prospects back in 1953.

Ironically, we were chortling with laughter only a year after the Bakersfield earthquake had rattled windows and destroyed buildings in that inland community less than one hundred fifty miles from our own coastal homes. Shortly after that episode I was taking a course in navigation at the U.S. Navy Quartermaster School in Bainbridge, Maryland. The chief petty officer teaching the course told his class how he once had to explain to his ship's captain how their vessel nearly ran aground on an uncharted island in the Pacific shipping lanes.

"It was really strange," the chief recalled. "This island popped up out of nowhere, then in a few months it was gone again."

He shrugged when we asked his opinion of what had really happened. "I don't know; one day it was wide, open water and the next day there's this barren, rocky, steaming island, and then, it vanished again in a few weeks. The only thing I'm sure about is that we were in the same spot of ocean, and that the piece of rock popped up and then submerged again."

Here we were, students of navigation, obviously intelligent human beings, talking about a fantastic natural phenomenon, but we really didn't let its significance sink in. The chief was not a geophysicist; he cared only that such emerging bodies of rock were a hazard to navigation. If memory serves me correctly, I cared only that I had a date with a Wave that evening. The point is, we take these signs of crustal instability for granted and don't give any real thought to the potentials for disaster.

When it comes to overpopulation and food production, we have attempted to counteract potential worldwide problems. But, our society has no contingencies for eventual cataclysm—we simply choose to believe such a catastrophe cannot occur, or we refuse to even consider the possibility at all.

I am suggesting that we reorganize the geological paradigm drastically and quickly, though I am reasonably certain a major cataclysm will occur before we can overcome our impotence in the changing of dogmatic notions.

Consider this: If the earth is not cooling but heating, then the stability of the continents is further threatened by pressures from beneath the crust. Most experts agree that the earth's crust is only twenty to forty miles thick, or as thin as an eggshell, relatively speaking. What it's like beneath that crust is only conjecture, but it is reasonable to assume that laws of physics apply.

Following is a statement made by geophysicist L. B. Slicter, who spoke to a group of chemists, geologists, and geophysicists attending a meeting of the National Academy of Science in 1950: ". . . the earth probably has grown by the accretion of relatively cool materials which were not molten at the outset. Our conceptions of the development of the primitive earth are, to say, the least, obscure. It is even uncertain whether the earth today is cooling or heating at depth, but the odds seem to favor the hypothesis of a heating earth. . . ."

Not much has changed since that speech was delivered. Geologists and other specialists have gathered more evidence from the sea bottoms and from Antarctica, but we're still saddled with suppositions and a myriad anomalies. It's quite obvious the subject is wide open for reasonable discussion, yet if someone comes forth with a thesis that concludes a cataclysmic rather than a slower geological evolution, there arises a hue and cry that such thinking is deranged.

If it weren't for some major improvements in social behavior, the likes of Immanuel Velikovsky, Charles Hapgood, and Erich von Daniken would be burned at the stake for challenging faded, old paradigms. All the

9

ancient-astronaut fans would be imprisoned for their own protection and one Tom Valentine would have his tongue yanked out.

Despite improved social behavior, the staunch defenders of old paradigms persist and still manage to arbitrarily put down challenges and hold sway over public opinion. For example, I telephoned the Universities of Chicago and Northwestern as I began writing this book to ask the geology professors their opinions about cataclysmic upheaval and violent movement of continents. Both learned men sneered at the concept, but neither took the time to elaborate on their own stance or to explain the reasoning behind their derision.

You won't get much argument that the continents on this planet have moved; it's how they moved and how fast that is debated. It's an old debate among geologists, and the key phrase in the disputes that have erupted during the past hundred or so years is "polar shift."

The problem is paleontologists and biologists are constantly finding fossils in the wrong places. Fossil corals in the Arctic Ocean, coal beds and fossil water lilies from the Arctic island of Spitzbergen, and other positive indications of warmer weather at the poles have been placed in evidence over the past two hundred years. What does it mean?

Let's casually analyze this notion of polar shift. Has the North Pole always been located in the water at the top of Canada and Alaska and Siberia, and the South Pole always on the Antarctic continent? Has this polar arrangement been stable since the planet was formed? These questions have bedeviled scientists for a long time.

Polar Shift can mean various things. It can mean a changing of the position of the earth's rotating axis in relation to the other heavenly bodies. You may have noticed that pictures of our solar system always depict our planet as being tilted in relation to the plane of the

10

sun's equator. Any change in the position of the earth's rotating axis would be very important to us. For example, if one of the poles pointed directly at the sun, we would have one very hot and sunny polar region and one very cold and dark polar region. A change in our tilt relative to the sun's equator is generally discredited by experts because the laws of physics tell us it would take some kind of major interplanetary collision to cause such a change.

Polar Shift can also mean a sudden relocation of the earth on its axis without changing position relative to the tilt from the solar orbit. This means the entire ball of real estate would careen and bring other points to the poles; the entire planet would therefore have to swivel around.

Again the laws of physics tell the experts that such a swivel motion is practically impossible. As the earth spins it is kept in equilibrium by the bulge at the equator which acts like the stabilizing rim of a gyroscope. The forces required to cause such a shift must overcome the stabilizing force of the equatorial bulge, and all our calculations, which work rather well for us in everyday engineering, tell us such a careening motion is impossible.

As you can see, an impasse was reached. Geologists were forced to invent theories to account for the fossils while keeping the poles in their proper place. Naturally common sense and the fossil evidence suffered, but every proposal for a polar-change theory was beaten back by the defenders of the paradigm.

Then came a great compromise, and today it is happily latched onto by eager traditional geologists who see it as a way to have warm-weather fossils in the Arctic Ocean, but without admitting cataclysmic continental movement. The compromise theory is called "continental drift."

In 1924 a scientist named Alfred von Wegener came up with the idea that the continents slowly drift around

11

on the globe, putting them in different positions relative to the poles at different times during the long geological history of the planet. His original thesis has been modified considerably and is called "plate tectonics" today. It is generally accepted now that the crust of the earth is made up of a dozen tectonic plates, which are outlined by "rings of fire," or volcanic and seismic activity. In the August issue of 1976, *Smithsonian Magazine,* a brilliant cartographer, Athelstan Spilhaus, shows how geometry and imaginative projections can make an exciting, visual case for a massive continent, called "Pangaea," separating slowly over millions of years into the various continents as they exist today.

Acceptance of the plate theory posed problems for geologists peering at photographs of Mars, and they became confused when they could see no signs of mountain building or movement of surface plates dictated by the forces of molten rock beneath the crust. Yet the Martian landscape featured volcanoes far larger than anything on earth. One such volcano, dubbed "Olympus Mons" by the experts, is so vast that its base would reach from New York City to Montreal, Canada. Although Mars is about half the size of the earth, there is a canyon on Mars that would stretch two-thirds of the way across the U.S.; and there is a plateau that measures more than 30,000 feet higher than the average Martian surface—geologists remarked that this is "higher than Mt. Everest!"

I found it amusing to read that the experts, who don't really know anything about the causes of mountain building and so forth on this planet, were making guesses about the crust of a planet they were seeing only through photographs. Carefully read the following paragraphs written by Walter Sullivan of the *New York Times*:

Yet Mars does not show the features that on earth testify to the constant *slow movements* of great rigid

12

sections, or plates, of its surface. It is these motions that, in the generally accepted theory of plate tectonics, are *believed* to have opened the ocean basins, moved the continents, and formed the long mountain ranges that parallel the present or past boundaries.

Such ranges include the Andes, Appalachians, Himalayas, and Urals. No such features are visible on Mars. Nor can geologists studying the photographs find anything comparable to the mid-ocean ridges where ocean basins on earth are being split apart by diverging plate motions. (Emphasis added.)

The experts figure that Mars had a different landscape than earth because of too little evidence of plate movement. Sullivan's news story continues:

There is a widely held *suspicion* among geologists that some form of upward flow in certain parts of the earth's interior is responsible for the plate motions. In the more *extreme version* of this *concept*, "plumes" of hot semimolten rock supposedly rise from close to the earth's core beneath such sites as Iceland.

This would account not only for the volcanic activity that has created Iceland, but for the spreading of plume material beneath the rigid top layer of the earth to push Europe and North America apart. (Emphasis added.)

In light of that September, 1976, *New York Times* report, we shall consider objections to the Wegener thesis. First of all, we have learned since 1924 that the ocean bottom is not plastic; it is rigid crust similar to dry land. In fact much of it was indeed once above the water's surface. Ocean bottoms have mountain ranges and canyons and valleys. This means that the continents we see above the water cannot very easily "drift" over the tops of submerged continents.

Another problem with the drift theory emerges when a serious effort is made to match up the east coast of South America with the west coast of Africa, although

13

advocates of Wegener's theory never tire of pointing out how well the two continents "seem to fit." Two geologists, however, actually spent several years covering more than 25,000 miles of coastline trying to match up the geology of the two coasts and finally concluded that they didn't mesh well enough to substantiate Wegener's theory.

Still another hitch in the drift thesis, especially in regard to the recent innovative hypothesis that "plumes" of molten matter squeezed up to separate North America from Europe, is the fact that radiocarbon-dating techniques show that the last ice age in North America ended only around 10,000 years ago. Wegener's timetable had it coming to an end more than 30,000 years ago. Thus, in order to match the drift dating with the last ice age, North America would had to have moved away from Europe at a rate of 1,500 feet per year in order to reach its present location. It simply isn't moving that fast, and there is no evidence of its having moved that fast in times of recorded chart making.

Ignoring objections to the drift theory, which include the catchy little fact that the plates also must move up and down as well as from side to side, today's experts have pretty much accepted the theory of plate tectonics. There was no getting away from it, I guess. Back in 1950, when the members of the British Association for the Advancement of Science called for a vote on the drift idea, the scholars rendered a split decision.

Of course there is always an intriguing amount of fact behind accepted theories, and indeed it appears that the earth's crust is made up of a dozen "plates."

The idea of "Pangaea" is also intriguing. Geologists looking at the moon (thanks to our space exploration we got a good view of the back side, too) noticed that the side facing earth is relatively smooth while the side away from earth is rough and raised. Mars features a similar geological pattern. Experts think that earth may also have featured this "lopsided" geology millions of years ago, but then something split the continents apart and they began drifting around. Advocates of this hypothesis stress that the Pacific side of our planet is largely oceanic while the opposite side is primarily continental. To that I say, "no deal"—there's just too much evidence that the Pacific continent, called Lemuria by many and Mu by some, was once entirely above the surface of the ocean.

For our present purposes we can accept the notion that there are such things as tectonic plates and that they do drift a bit, but the idea of polar shift in relation to those plates cannot be suppressed forever because the evidence for it is overwhelming. Geologists have done some strange computing to account for the overwhelming body of evidence. One geophysicist suggested that since the continents cannot drift, perhaps they could "creep." Another thought that the wobble in the earth's spin could cause a plastic readjustment that might move the crust a quarter-turn around the globe in a million years. Several experts have contemplated collisions with planetoids forcing sudden continental movement, and you have just seen how many of today's geologists accept the idea of a plume of hot lava coming up from the center of the earth to squeeze continents apart. Nevertheless the paradigm of a substantially solid earth withstands and most people think "cataclysm" is a class Catholic kids attend on Saturdays!

In more recent times detailed studies of magnetic fields around the earth have provided more evidence to support the thesis that the poles have been in different positions relative to the continental masses.

The only *logical* solution to the problems presented by the evidence is the movement or shift of the earth's crust at various periods in the past. This is not a new concept, but rather one that has been hooted and

15

suppressed for many years. In his book, *Earth's Shifting Crust*, published in 1958, Charles H. Hapgood presents a substantial, thorough, and reasonable case for the displacement of the earth's crust relative to the polar regions, which indeed remain fixed relative to the heavens.

Because he challenged a dogmatic paradigm, Hapgood was fully aware of how his published work would be attacked, so you will notice the great care he used to differentiate between known fact and hypothesis as he wrote. Hapgood summarized the idea of crustal shifting as follows:

To understand what is involved in the idea of a movement or displacement of the entire crust of the earth, certain facts about the earth must be understood. The crust is very thin. Estimates of its thickness range from a minimum of about twenty to a maximum of about forty miles. The crust is made of comparatively rigid, crystalline rock, but it is fractured in many places and does not have great strength. Immediately under the crust is a layer that is thought to be extremely weak because it is, presumably, too hot to crystallize. Moreover, it is thought that pressure at that depth renders the rock extremely plastic, so that it will yield easily to pressures. The rock at that depth is supposed to have a high viscosity; that is, it is fluid but very stiff, as tar may be. It is known that a viscous material will yield easily to a comparatively slight pressure exerted over a long period of time, even though it may act as a solid when subjected to a sudden pressure, such as an earthquake wave. If a gentle push is exerted horizontally on the earth's crust, to shove it in a given direction, and if the push is maintained steadily for a long time, it is highly probable that the crust will be displaced over this plastic and viscous lower layer. The crust, in this case, will move as a single unit, the whole crust at the same time. This idea has nothing whatever to do with

16

the much discussed theory of drifting continents, according to which the continents drifted separately, in different directions . . . .

Let us visualize briefly the consequences of a displacement of the whole crustal shell of the earth. First, there will be the changes in latitude. Places on the earth's surface will change their distances from the equator. Some will be shifted nearer the equator and others farther away. Points on the opposite sides of the earth will move in opposite directions. For example, if New York should be moved 2,000 miles south, the Indian Ocean, diametrically opposite, would have to be shifted 2,000 miles north. All points on the earth's surface will not move an equal distance, however. To visualize this, the reader need only take a globe, mounted on its stand, and set it in rotation. He will see that while a point on its equator is moving fast, the points nearest the poles are moving slowly. In a given time, a point near the equator moves much farther than one near a pole. So, in a displacement of the crust, there is a meridian around the earth that represents the direction of the movement, and points on this circle will be moved the farthest. Two points, 90 degrees away from this line will represent the "pivot points" of the movement. All other points will be displaced proportionally to their distances from this meridian. Naturally, climatic changes will be more or less proportionate to changes in latitude, and because areas on opposite sides of the globe will be moving in opposite directions, some areas will be getting colder while others get hotter; some will be undergoing radical changes of climate, some mild changes of climate, and some no changes at all.

In that summary, Hapgood described only the changes in weather that would naturally be experienced when one part of the world moves to another location. There is much more to think about if such a cataclysmic shift of the tectonic plates occurs, a topic that will be explored in depth in later chapters. At this point it is important to know that Hapgood did not claim to be coming up with any new ideas. He was rehashing the work of many brilliant challengers to the stodgy paradigm of "uniformitarianism" laid down by the father of modern geology, Sir Charles Lyell back in the early 1880s. Uniformitarianism means that the geological processes of mountain building, canyon digging, glacier gouging, wind erosion and water erosion have always worked at the agonizingly slow place at which they are progressing today. The only new wrinkle Hapgood added was the mathematical wizardry of James H. Campbell, an ingenious engineer who computed how certain natural forces could indeed overcome the gyroscopic stability of our spinning planet and cause a sudden crustal shift.

Hapgood was also more scientific and detailed in his analyses of the evidence than other challengers, and I will borrow heavily from him in the geological portions of this book.

We are concentrating on all this geology for a reason. In the ensuing chapters, a case will be made for regularly occurring cataclysms in the earth, the swallowing up of past great civilizations, and the chances of such a thing happening to us. In his summary quoted above, Hapgood detailed the climatic changes, but such changes will effect only the survivors of a major continental drift. Imagine the seismic shocks (earthquakes) and *tsunami* (Japanese word for shock wave in water) that will occur! Monitors should pop their bolts and nuts with the Richter scale reading "impossible," or more than ten. Flooding of an unimaginable magnitude should occur. Cataclysms will rearrange much more than latitude. Entire continents can submerge and others emerge from the depths of the oceans which cover most of our planet. In the photo section are illustrations of all the movements now accepted by seismologists regarding "techtonics" or continent building and moving. The illustrations are taken from *Scientific American*, November, 1971. The article on the well-known San Andreas fault in California was written by Don L. Anderson, and it incorporated the latest thinking of scientists at the time. According to the seismic evidence a continental plate, or the earth's crust, can be compressed and stretched, folded, thrust upward, dropped downward, thickened and thinned. However, for some strange reason our scientists believe all this maneuvering is accomplished by very slow movement over millions of years. That just doesn't add up. Earthquakes are sudden, volcanoes are sudden when they really blow their tops, and there is every reason to believe that sudden shifting can also occur.

There will be a tremendous loss of life, but millions of people as well as animals and plants will survive, even though billions may perish. The very thought of such a cataclysmic disaster is so horrifying that it could well be one reason why traditional geologists would rather not even consider it possible. However, such violent events have occurred, and legend hints that it may have worked out for the better so far as mankind is concerned. ·

Yes, there will be extinctions of some species of animals, In fact cataclysm is a far more logical reason for the great extinctions of species of the distant past, such as dinosaurs and mammoths, than the rationalizations drummed up by uniformitarians. That's where we go from here. The next step is to reevaluate our prehistoric paradigm and its present irrational connection with our historic paradigm.

We have all been taught in school that such places as Atlantis and Lemuria are mere myths. We have been taught that mankind lived in a stone age for nearly a million years until suddenly, just a few thousand years ago, he got smarter. We are constantly taught that we can learn from history, that we can improve on our past mistakes. Sometimes this may be

19

true, but if we are going to learn anything valuable from our past, we had better make sure our view of the past is correct and not merely construed to fit a comfortable paradigm.

CHAPTER 2

The Ice Ages:
A Geological and Educational Hoax

Ice is nice. We love it in our drinks, we need it to keep fish and dairy products fresh, and there's no finer skating surface. Ice plays an important role in our lives—in fact it plays a far more important role than had heretofore been imagined.

All of us have learned at least something about the great ice ages that evidently existed thousands, or in some cases, millions of years ago. Grammar school kids often thought their teachers had to be a little nutty to talk of glaciers in humid Indianapolis, southern Ohio, and even North Carolina. But, the science of geology assures us, such huge masses of ice indeed covered most of the world at one time or another, and evidence indicates that the ice coverage extended through much of North America much as it covers Antarctica today. How could that be? How could such a chilling thing happen? We are told that glaciers were

21

gouging out the Great Lakes only a scant 10,000 years ago!

To those of us who have been taught to think in terms of eons past for the ice ages, the latter is a shocking statement, but it really poses some problems for the uniformitarian geologists. As we have already seen, enough evidence has piled up in recent years to force the uniformitarians to at least accept that continents can drift. However, even with creeping continents, the explanations we laymen are handed for such ice ages are more than a bit fanciful.

You can imagine the difficulties the early geologists had when they uncovered the overwhelming evidence that vast sheets of ice really did cover most of the temperate and even some of the tropical zones. Surely the people laughed when the first geologists suggested that the only way certain formations of rock and other debris could have been formed was by the weight of glacial ice. It took a long time, but finally the evidence for ice sheets caught on, was accepted, and we were presented with a new paradigm.

Curiously, no other field of geology offers more evidence for sudden displacements of the earth's crust than glacial ice studies, yet it remains the classical geologists who most bitterly oppose any notion of cataclysmic change. Geologists have worked their brains to the bone trying to come up with substantial explanations of how glaciers could grow and cover vast areas in locations with notably warmer climates. But the more modern science learned about climatology and meteorology, the more impossible the ice age theories became.

We are supposed to accept that, like Topsy, the ice just grew and grew. In his book, Hapgood seemed to have some fun quoting the experts on the problem. However, listening in on a junior high school teacher telling students about ice ages back in the times of mastodon and woolly mammoths would prompt one to conclude there was never any doubt that somehow the climatic conditions on this planet changed drastically and ice crept down on North America from the polar region, which was where it is today.

Ancient ice caps have refused to cooperate with the position of the poles as we know them today, but the experts still toiled long hours in order to explain how the ice fanned out from the poles and covered warmer areas. In 1929, geologist A. P. Coleman wrote:

In early times it was supposed that during the glacial period a vast ice cap radiated from the North Pole, extending varying distances southward over seas and continents. It was presently found, however, that some northern countries were never covered by ice, and that in reality there were several more or less distinct ice sheets starting from local centers, and expanding in all directions, north as well as east and west and south. It was found, too, that these ice sheets were distributed in what seemed a capricious manner. Siberia, now including some of the coldest parts of the world, was *not* covered, and the same was true of Alaska, and the Yukon Territory in Canada; while northern Europe, with its relatively mild climate, was buried under ice as far south as London and Berlin; and most of Canada and the United States were covered, the ice reaching as far south as Cincinnati in the Mississippi Valley . . . .

Now Coleman didn't like cataclysmic changes as a hypothesis any better than the next geologist, but he confronted the evidence face-to-face. Careful geological studies determined that ice, which gets pretty heavy when it's a mile thick, was responsible for chewing up vast chunks of real estate in different parts of the world during different periods of time. For example, way back in the Permo-Carboniferous era, glaciation took place in areas that are now southern temperate zones, and yet, during the same period there is no evidence

23

that glaciation took place in present day polar regions. Then there is the case of India back in the Carboniferous era (geologists deal in millions of years like you and I deal in weeks; it is important only to know that this was a long, long time ago). It seems that the ice sheets covering India moved from *south* to *north*! Egads! You can imagine how this bit of evidence plays havoc with the "fanning out from the poles" theories that our kids still learn in school. Either India was much closer to the South Pole at the time, or the weather was lousier than can be believed.

Dr. Coleman did a great deal of his ice age research in the stifling heat of Africa and India. He commented, "The dry, wilting glare and perspiration made the thought of an ice sheet thousands of feet thick at that very spot most incredible, but most alluring."

Traditionally we are told that ice ages occurred as a result of the lowering of temperatures across the entire planet at the same time. No, this isn't a joke—this ridiculous assumption is the one generally accepted. Even though such an idea is in sharp conflict with known meteorology and physics, it has been in vogue for at least 100 years. Let's look at the conflicts that stem from our accepted view of ice ages:

1. The evidence shows that during the ice ages the areas outside of the ice caps were drenched with far more rainfall than is experienced in most parts of the world today.
2. Together with the ice caps themselves, this means there was a much, much higher average rate of precipitation.
3. To account for such heavy rainfall and snowfall there must be a tremendous supply of moisture in the air.
4. The only possible way to get so much moisture into the air is to *increase* the temperature of the air.

Thus, ice ages require that our atmosphere heat up, not cool off! In fact, according to the physics of evaporation and precipitation, it takes one helluva lot of heat to get up enough moisture to make a glacier. Therefore, the only thing that might account for the phenomenon at all is a violent rending of the earth's crust, which creates massive volcanic upheaval while concurrently thrusting fresh new territory into a polar region.

In order to account for a worldwide lowering of temperatures, the experts of this absurdity have gone to great lengths—in fact, every now and then we read in our own newspapers that one scientist or another feels we are on the verge of "entering into another ice age."

Here are some of the assumptions used to explain worldwide temperature fluctuations of a magnitude that would put glaciers in the Congo River basin:

1. Variations in the quality and quantity of particles the sun emits.
2. Outer space dust or gas clouds getting in the way and keeping the warming rays from hitting our beaches.
3. Very cold particles from outer space entering the earth's atmosphere.
4. Huge clouds of dust in our atmosphere and a higher proportion of carbon dioxide in the air, both due to volcanic eruptions.

There is no evidence to support assumptions two and three, and assumption one fails because ice ages last for exceptionally long periods of time while sunshine variations last only short periods of time. Assumption number four fits well with what we can expect to occur with the next cataclysm, and we'll delve further into this interesting meteorological concept in a later chapter.

If ice ages cannot be explained by assumptions about interference with sunlight, how about using changes in the relative positions of the earth and the sun? Elaborate theories involving the earth's orbit and the wobble in the earth's rotation that causes "precession" have been put forth, but they are just as weak as the "sunshine block" theories. For example, if the temperature of the entire planet cooled enough for glaciers a mile thick to form at the equator, how was life sustained?

I hope that by now you are saying, "My God, I didn't know all this." If so, it is because you haven't given the subject any thought for the simple reason that ice ages are not really relevant to everyday life. At least they aren't yet!

But there is much more to the myth of the ice ages all of which will play a key role in our presentation.

Time means nothing to the geologist—he'd just as soon deal in 150,000-year blocks of time as he would 12,000. After all, what's a billion or two trips around the sun so long as they happened before today? The modern scientific technique of radiocarbon dating tells us that the last ice age (called "Wisconsin glaciation" by the specialists) ended about 10,000 years ago, not 30,000 as previously supposed.

Those brilliant men whose work was distorted into the atomic bombs of World War II learned a great deal about nuclear reactions, and they made possible the development of various techniques for radio-isotope dating—including the fairly well-known radiocarbon, or carbon 14, dating method. Willard F. Libby of the University of Chicago developed carbon 14 dating techniques. Carbon 14 is an isotope of carbon that has a distinctive "half-life" period of around 5,500 years. Since radiocarbon exists in nature and has a relatively short half-life, or period when it loses half its mass by radiation, the amount of it in any substance containing organic carbon will decline perceptibly over a few centuries. By determining how much carbon was contained in the original specimen and then measuring what remains, a date can be assumed within only a small margin of error.

In 1951 radiocarbon-dating techniques told us that the Wisconsin glaciation was *still advancing* as recently as 11,000 years ago. This means that about 4 million square miles of ice melted in only a couple of thousand years. Just imagine the extent of flooding; Noah, along with his ark, comes to mind. Not only does this prove that traditional geologists were all washed up when they figured the ice age time span, but it clearly wipes out any assumptions that slow, cosmically caused weather changes, or "creeping tectonic plates" were responsible for the demise of the ice cap.

As radiocarbon dating improved, the ice age in North America became a traditional geologists' nightmare. The period of 150,000 years formerly estimated for the Ohio glaciation was cut back to 25,000 years. Dr. Leland Horberg, an eminent geologist who did most of the primary research evaluations in 1954, actually wrote that the evidence was "appalling" from the standpoint of accepted geology. He suggested that either the concepts of gradual change be abandoned, or that radiocarbon-dating methods be questioned. So rather than change the paradigm, the reliability of dating was challenged, then the issue was simply allowed to die out without any widely published conclusion. Horberg himself wrote: "Probably only time and the progress of future studies can tell whether we cling too tenaciously to the uniformitarian principle in our *unwillingness* to accept fully the *rapid* glacier fluctuations evidenced by radiocarbon dating." (Emphasis added.)

The implications of a massive sheet of ice, just like the one on top of Antarctica today, developing and then vanishing within a period of only 25,000 years are staggering to the imagination. The only likely cause of such an event is a drastic displacement of the earth's crust and cataclysmic upheaval thrusting parts of North America into the polar region. And even though we're

27

getting ahead of ourselves, this clearly indicates that such "mythological" places as Lemuria in the Pacific and Atlantis in the Atlantic not only could have existed, but could well have flourished with civilizations.

For the last great ice age in America to make any sense at all, the assumptions must account for the rapid fluctuations. If this "recent" ice age came and went in the "twinkling of an eye" geologically speaking, then we must assume other, more ancient ice ages also did the same; it doesn't make much sense to assume uniformitarianism for one ice age and cataclysm for another. Again, the only substantial answer is rapid and violent crust displacement.

Hapgood concluded rather firmly that: "None of the great glaciations of the past can be explained by the theories hitherto advanced. The only ice age that is adequately explained is the present ice age in Antarctica. This is excellently explained. It exists, quite obviously, because Antarctica is at the pole, and for no other reason. No variation of the sun's heat, no galactic dust, no volcanism, no subcrustal currents, and no arrangements of land elevations or sea currents account for the facts."

As I said at the beginning of this chapter, ice will play a major role in our daily lives, specifically the tremendous ice cap covering Antarctica. We are going to look at it from two points of view. First, let's look at the evidence showing that this frigid continent was not always so polar, and later on we will see how the tremendous weight of this ice has an effect on the planet's rotational spin and, thus, will play a key role in the next cataclysm.

The polar continent was much warmer in times past, as testified to by another form of nuclear dating. Just as radiocarbon dating jolted the North American ice age paradigm, the radioelement inequilibrium method jolted the accepted notions of the Antarctic ice age. Popularly known as the "ionium" method, this dating

28

system works well with sea sediments. The method is based upon three radioactive elements: uranium, ionium, and radium. All three are found in sea water and in sea sediments, and they decay at different rates. This decay variation means that different proportions of these elements will appear in a sample of sediment at different times. The ionium method permits dating back as far as 300,000 years.

Oceanographic surveys take core samples of sea sediment for analysis. In 1948 during Admiral Byrd's expeditions to the North Pole, core samples of the sediment on the bottom of the Ross Sea were taken. Surprise, surprise! Among the anticipated sediments of glacial types were found strata of fine sediment of the kind deposited by rivers in a temperate climate. Ionium-dating techniques tell us that during the past million years Antarctica has periodically enjoyed a moderate climate. Now what caused that?

Moreover, these drastic changes in climate took place at relatively rapid intervals; the last ice age in Antarctica began only a few thousand years ago—about the same time the Wisconsin Glaciation vanished.

Even more amazing is the correlation of these geological facts with historic document of unknown origin, known as the "Piri Reis maps." In the 1500s a Turkish geographer, Piri Reis, compiled maps from ancient geographic projections that may well stem from the earliest Mediterranean civilizations. For years the Piri Reis projections baffled scientists, including the famed explorer, N. E. A. Nordenskjold, who spent at least 17 years attempting to understand them. Maps are not simple things, and how certain maps are projected can be extremely complex. Finally an archeologist and cartographer, Arlington H. Mallery, solved the projections puzzle in the mid fifties. We now know that the map indicates all the coasts of South America and a great portion of the coast of Antarctica. This map of unknown age, but assumed to be more than 5,000 years old, showed the topographic features of Queen Maud's Land and Palmer peninsula and had obviously been projected when Antarctica was ice-free. Today seismic technology tells us that the topography of the land beneath all that ice corresponds precisely with the ancient map.

Quite a lot of fuss has been made about how the Piri Reis maps most certainly had to be made by people with aircraft, and that's something worth fussing about since it violates our present historical paradigm, but the warmer climate in Antarctica just 10,000 or so years ago correlating with civilized man tells just as exciting a story. We are just beginning to purge our faulty paradigms. The defenders seem to be battling to the bitter end, but if reason prevails, we shall all have a new view of prehistory, history, and our future.

Detractors with excellent credentials will point out that there definitely was a prolonged period of high temperatures following the last ice age in North America. Naturally there are annual fluctuations in weather, but for a period of about 2,000 years, beginning about 5,000 years ago, our planet had what climatologists call the "climatic optimum." That outstanding weather fact has been credited with the so-called "last warm period in Antarctica." However, the dating techniques, now well established as accurate, indicate the last "warm period" for Antarctica began at least 15,000 years ago.

So much for the "myth" of ice ages. As we shall see, these periods of tremendous change are *not* ice ages but are instead "between cataclysm ages." For example, Antarctica is experiencing an ice age, but only because the earth overall is "between cataclysm ages." Not long ago, relatively speaking, the North Pole was located in Hudson Bay and we had ice covering the upper Mississippi Valley. Antarctica was not centered at the South Pole and the huge plains of Siberia were much warmer than they are today. There was a large land mass in the center of what is now the Atlantic Ocean and a highly developed civilization lived there—Atlantis! Then one day, probably between October 31 and November 17, the earth shuddered, its crust shifted a few thousand miles along a meridian in the western hemisphere, and Atlantis sank into the sea. Meanwhile Wisconsin slipped southward to its present location and ice started melting at a fantastic rate. Survivors struggled against the elements, soon forgetting much of their culture's technology, but remembering just enough to create interesting "legends," such as the "coming of the Angels of death" on Halloween, celebrated by all cultures sometime between October 31 and November 17.

The details of that relatively minor cataclysm will enter into the discussion again in a later chapter. Here it is deemed "minor" only in comparison with the cataclysm that occurred about 30,000 years ago and sunk the huge continent of Mu, or Lemuria, beneath the present day Pacific Ocean.

Those two cataclysms, and other lesser isostatic readjustments in between, accounted not only for the extinction of two advanced civilizations and the hot-and-cold growth/shrinkage of the ice caps, but for millions upon millions of animals, some of whose remains make up an impressive body of evidence for such cataclysms occurring. Enter woolly mammoths, et al!

The woolly mammoth is a large elephantine creature which is invariably pictured standing near a glacier to convey the idea that it was indeed an ice age resident. Schoolchildren are misled into believing that this huge mammal was hunted by Neanderthal-like cave men with stone-tipped spears, thereby signifying that the mammoth existed long before man became civilized. Then, apparently, the cave men and other natural forces failed to stick to their hunting license limits and for "no apparent reason" the mammoths and others became extinct.

Bunk! The reason for their extinction is as plain as the nose on Jimmy Durante's face. Anything so violent as a cataclysmic shift of the earth's crust is going to be accompanied by tremendous storms, floods, upheavals, volcanoes, quick freezing and rapid thawing on an unbelievable scale. Violent death came for millions of creatures, including human beings, the last time the earth shuddered. Among the victims, the most romantic seem to be the huge mammoths whose remains have been occasionally unearthed for the world to ponder. The most famous of all these mammoth carcasses is the so-called "Bereskova" mammoth found along the banks of the Bereskova River in frigid Siberia back in 1899.

The literature on this particular mammoth is terribly sensationalized and overdone, and I would avoid it except that it helps make a point. Cataclysmologists have used the Bereskova mammoth to "prove" a cataclysm did indeed occur, although, of course, no such thing can be proved by one mammoth carcass. On the other hand, the experts whose writings only appear in the technical publications have made some glaring booboos, too.

In 1960 Ivan Sanderson wrote an article for the *Saturday Evening Post* elaborating on the "riddle of the frozen giants." The article was brilliantly done and posed many of the same questions I'll pose later in this chapter. However, the scientific community failed to appreciate Sanderson's conclusions that cited violent cataclysms as cause for the extinctions, and Dr. William R. Farrand wrote a lengthy rebuttal for *Science* in March, 1961. Following is what the scientist had to say about the popular scientist-science fiction writer:

In contrast to scientific efforts, a number of popular and quasi-scientific articles have appeared in recent years, in which fragmentary knowledge, folk tales, and science fiction are combined under the guise of verity—much to the chagrin of scientists and the confusion of the public. The most recent of such articles is that of Sanderson, who comes to the conclusion that the "frozen giants" must have become deep-frozen within only a few hours' time. Such a thesis, however, consistently disregards the actual observations of scientists and explorers. Adding insult to injury, Sanderson proceeds to fashion a fantastic climatic catastrophe to explain his conclusions.

The first thing Farrand does to debunk the cataclysmic approach is point out that his article will *not* deal with other species of mammoths—only the woolly mammoths. This, of course, implies that other species of the critter didn't die off because of a cataclysm, even though the author does not specifically state his assumption.

Typical of the dogmatic approach. Farrand stresses that "the habitat of the woolly mammoth is indicated clearly by its physical appearance and food habits, as determined from the frozen carcasses and associated fossils. Long hair, thick wool, and a heavy layer of fat definitely indicate a cold climate."

Sounds "scientific" doesn't it? Not in my book! Let's analyze what was just stated carefully. None of the features of the woolly mammoth—thick skin, hairy coat, thick wool, and a layer of fat—necessarily indicates an arctic climate. Detailed studies show that the hide of the woolly mammoth and that of the Indian elephant (tropical climate) are almost identical. It was also shown that the mammoths lacked sebaceous or oil glands within the skin, a vital item for surviving winters in cold climates. Thick fur or hairiness doesn't mean adaptability to cold—tigers have thick fur. As for a layer of fat being used to protect against the cold, this

33

is not verified by physiologists. Fat attests to an abundance of food—something not necessarily associated with the ice ages. Layers of fat generally mean the animal is storing food. Camels, for example, store more food than most animals and they most assuredly don't live in a cold climate.

Much to the chagrin of this layman, Farrand's conclusion about the mammoth being an arctic mammal is not borne out by scientific scrutiny.

Then there is the loud suggestion by cataclysmologists that the Bereskova mammoth was suddenly slammed into a mud bank with terrific force while he was complacently eating "buttercups," which they claim are a warm climate flora. These statements have been exaggerated by overenthusiastic cataclysm freaks.

The contents of the Bereskova mammoth's tummy, which remained well perserved all these thousands of years, is of particular interest to everyone because it obviously proves the type of flora existing around the creature at the time it was alive. Without going into the dull botanical details, and without sensationalizing, here is what the Russian experts determined made up the contents of the Bereskova mammoth's stomach:

1. No remains of conifers were found. Therefore, the assumption that the woolly mammoths fed mainly on coniferous vegetation put forth for years by experts is erroneous.
2. Several kinds of grasses were found, all of which could indeed be found in Siberia today, and this includes buttercups.
3. Only one of the plants found in abundance in the beast's stomach fails to fit the arctic format—*Agropyrum cristatum* L. Bess, which is a temperate, plains flora.

So despite all the hullaballoo, there isn't a great deal to get excited about. One small item of great interest is the fact that the mammoth was lying in vegetation totally different from what it had been eating!

Most advocates for cataclysm as the cause imply that the Bereskova mammoth was "obviously" killed by the violent forces and then "quick frozen" when the formerly temperate prairie shifted into the polar region. The hypothesis is not all bad, but it's not strong enough to withstand serious scrutiny. The animal had some broken bones, but judging from its stance when found, it more likely received them from a fall into a crevice than from being "slammed" by an earthquake. It was a bull mammoth and had an erect penis, indicating suffocation or asphyxiation as the probable cause of death. It was quite possibly buried alive.

Stories of how fresh and edible the meat of this mammoth was when it was unearthed are greatly exaggerated. The carcass was amazingly well preserved for such a long period of time, but putrefaction had indeed set in prior to the original freezing, and the stench as it thawed was described as "unbearable" by the excavators. Some of the meat was marbled and fibrous, but only dogs showed any appetite for it. All stories about a feast on mammoth meat are merely colorful fiction.

Farrand concludes that "although some problems concerning the frozen fauna of Siberia and Alaska remain to be solved, there is no reason to assume the occurrence of catastrophe or cataclysm." He stresses that only thirty-nine carcasses have been unearthed, only four of which were complete, and claims this certainly does not indicate a violent, massive slaughter. And in this part of the assumption he is correct—the Bereskova mammoth probably missed the cataclysm. However, the final lines in Farrand's article, which he probably felt settled the issue for all "serious scientists," read: ". . . it is very unlikely that a catastrophic coagulation occurred in Siberia. On the contrary, the frozen giants are indicative of a normal and expected (uniformitarian) circumstance of life on the tundra."

He conveniently omits any mention of the zillions of ripped apart, schredded, twisted, and torn pieces of animal bodies also found in considerable abundance in the vast "arctic muck."

In 1946, Professor Frank C. Hibben laid the evidence on the line with his excellent firsthand accounts in *The Lost Americans*. The evidence of sudden, violent death on a massive scale lies in the arctic muck, sometimes more than 4,000 feet below the surface. He pointed out that "trainload lots" of animal bones and bits pack the frozen, dark gray sands that cover as much as one-seventh the earth's land surface, or nearly all of that portion ringing the arctic sea.

He listed the various kinds of animals whose remains clutter the permafrost: "Mammoth, mastodon, several kinds of bison, horses, wolves, bears, and lions. . . ." Contrary to the low-keyed assessment by Farrand, Hibben writes colorfully about what the geologist encounters in the frozen north:

> Within this mass, frozen solid, lie the twisted parts of animals and trees intermingled with the lenses of ice and layers of peat and mosses. It looks as though in the midst of some cataclysmic catastrophe of 10,-000 years ago the whole Alaskan world of living animals and plants was suddenly frozen in motion in a grim charade. . . .
>
> Throughout the Yukon and its tributaries, the gnawing currents of the river had eaten into many a frozen bank of muck to reveal bones and tusks of these animals protruding from all levels. Whole gravel bars in the muddy river were formed of the jumbled fragments of animal remains.

All those animals didn't die of old age. The signs of violence are unmistakable. Hibben also makes it clear that the ice sheets were not the cause of death. He writes:

Their bones lie bleaching on the sands of Florida and in the gravels of New Jersey. They weather out of the dry terraces of Texas and protrude from the sticky ooze of the tar pits of Wilshire Boulevard in Los Angeles. Thousands of these remains have been encountered in Mexico and even in South America. The bodies lie as articulated skeletons revealed by dust storms, or as isolated bones and fragments in ditches or canals. The bodies of the victims are everywhere in evidence.

To further illustrate the hold the traditional geological paradigm has on its experts, we note Hibben's wording in the following paragraph:

One of the most interesting of the theories of the Pleistocene end is that which explains this ancient tragedy by world-wide, earth-shaking volcanic eruptions of catastrophic violence. This *bizarre* idea, *queerly* enough, has considerable support, especially in the Alaskan and Siberian regions. . . . (Emphasis added.)

Obviously Hibben didn't want to convey the impression that he was in any way, shape, or form a "weirdo" cataclysmologist.

Along this same line of thought, many cataclysmologists have been accused of sensationalizing the evidence of the frozen tundra of Siberia and Alaska. For a slight change of pace, here is an excerpt from a very conservative geological report taken from the "Geological Society of American Bulletin," published in 1943. The author is Stephen Taber who made the observations firsthand, but draws no conclusions about related causes; draw your own conclusions:

The distribution of fossil trees also indicated a somewhat milder climate during the silt accumulation than that now prevailing. Today Seward Peninsula is treeless, except for a thin growth of small spruce in

37

some of the valleys in the extreme eastern part, but large spruce logs are found in the silts at many places on the peninsula and even farther north. It has been suggested that the logs floated from distant sources during a period of high sea level, but the silts unquestionably accumulated under subaerial conditions for they contain numerous bones of land mammals, small twigs, spruce cones, and even tree stumps in place. In the silts of Candle Creek and its tributaries, birch and willow logs six to ten inches in diameter, as well as spruce, occurred with the bones of mammoth, bison, and other Pleistocene animals. On Quartz Creek bones of mammoth and horse were found with the trunks of large spruce trees, one of which is said to be five feet in diameter and eighty feet long. Moffit and others have also reported large spruce logs in the valley silts of Seward Peninsula. At Elephant Point, Quakenbush found a tree stump in place. Spruce logs, including one with roots attached, were found along the Ikpikpuk in Lat. 70°N. at an elevation of more than 200 feet although spruce is not known to live north of the Brooks Range.

In Lat. 74°N. on Great Lyakhov Island in the New Siberian Islands group, Toll found specimens of alder (*Alnus fruticosa*) "consisting of the whole trunk and roots to a length of fifteen to twenty feet," with bark intact, although its northern limit today is 4° to the south. Willow and birch, bearing well-preserved leaves, were also found.

Taber's report is certainly not sensationalized, and yet the evidence undeniably points to forests of trees existing a long time ago in a part of the world where they don't exist today, and something suddenly came along to smash them, then "silted" them, then "froze" them. The geologist was obviously competent at categorizing the evidence and in his later report, regarding the demise of millions of animals, his confusion as to the causes becomes fairly apparent:

38

Fossil bones are astonishingly abundant in the frozen ground of Alaska, but articulated bones are scarce, and complete skeletons, except for rodents that died in their burrows, are almost unknown. Of several tons of bones of the larger mammals seen in 1935, four bison vertebrae were the only ones found in a position indicating original articulation. The dispersal of the bones is as striking as their abundance and indicates general destruction of soft parts prior to burial. However, occasional articulated bones indicate primary burials. . . .

Numerous references are found, especially in semipopular literature, to "bone yards," "bone pits," "bone beds," "elephant graveyards," and natural "traps," but the writer has seen no evidence of such features. The bones are naturally more abundant in some localities than others, but they have been found in practically every valley where placer mining has been carried on.

The abundance of fossil bones in the nonglaciated area has been attributed to northward crowding of Pleistocene animals by an expanding ice cap, but the decrease in the habitable area took place gradually, and at any given time the number of animals per unit area was determined by the food supply and similar environmental factors. It has not been proved that the deposition of the silt coincided with the period of maximum ice expansion, but, even if it did, there is no reason to assume that the density of the animal population would have been greater north of the ice cap than south of it, where the climate was milder. The abundance of both plant and animal fossils in the Alaska silts, as compared with their occurrence in Pleistocene deposits south of the glaciated area, must be attributed to better preservation in the colder climate.

The facts then tell us a different story than that related by traditional geology. Not only can we cite evidence that some gigantic force appearing with sudden ferocity destroyed millions of animals and forests, but the very act of large tracts of land being suddenly thrust into polar climates has been preserved for the record beneath the frozen tundra. Traditionally the arguments against such cataclysmic change have stated that there is no evidence in other parts of the world to support such a premise. What ignorance to come with so much arrogance! The bacteria in warmer climates have done their job on the evidence; it's been recycled—it was biologically degradable. In my opinion, uniformitarian geology must now take its place alongside the other discarded paradigms of history.

Countless cataclysms have occurred during the earth's lifetime, and we must assume they will continue to occur so long as we have a planet with a rigid crust riding atop a layer of molten magma, which is constantly heating and expanding and forcing itself upward against the continental plates.

Once the cobwebs of uniformitarianism are cleared from our heads we can look at the mechanics of cataclysm with open minds, without sneering at how "impossible" it is.

CHAPTER 3

*This Flying Pressure Cooker
We Call Earth*

Nobody really knows what's going on in the center of the earth—not even Jules Verne. The best we can do to figure out what it is like beneath our crust is to theorize using what we think we know about the physics of a spinning mass as a basis. Since volcanoes do erupt from time to time, we know that there are a lot of hot rocks held under pressure somehow. It is generally assumed the earth has a fiery core, a huge mantle, then a layer of molten matter, and finally a thin, hard crust. We won't argue with that view because it stands up pretty well to the facts. (See photo section for illustration.)

The National Geographic Society, in advertising their latest book, *Our Continent*, describes the makeup of the earth with an excellent analogy: "Compare our globe to a soft-boiled egg: thin, shell-like crust, thick middle layer representing the white, and a partly liquid, but fiery core equivalent to the yolk."

We can assume that *National Geographic* used only the most recent, authoritative sources in putting its new book together, so that analogy is most likely very sound. But I'm going to pick at the information the people at *National Geographic* are putting out in *Our Continent*, only because I'm not yet ready to get off the back of puffed-up authoritarianism.

They fully support the "Pangaea" thesis that has already been mentioned, and they like the idea of "drift" for the tectonic plates. However, they credit cataclysm only "in the cataclysmic beginnings" of the planet. They ignore tremendous amounts of evidence not only for cataclysms, but evidence that is *opposed* to some of their accepted notions—such as that of the Americas matching up well with Africa. They write: "Bits of Georgia remained in Africa when the continents—once a single land mass—split." They don't point out that this is only theory (and not really a substantial one) and that more than 25,000 miles of the two coastlines do not mesh well.

They tell us that the "Atlantic grows wider by about an inch a year . . . and parts of the Pacific spread four times as fast. . . ." Now that statement is factual, based upon oceanographic surveys, but it isn't necessarily truthful. First of all it is slightly sensational to use the phrase "four times as fast" to describe a rate of four inches per year, especially when measuring something as gigantic as the Pacific Ocean. We've already stressed how the slow movement of the Atlantic rift does not jibe with radiocarbon dating of the last ice age. Something is out of proportion in the way knowledge is handed to us. Hapgood was mercilessly attacked for his excellent book by the "authorities," but it is perfectly all right for the same "authorities" to put forth unsubstantiated guesswork.

Despite my nitpicking skepticism, I heartily recommend the *National Geographic* book as fascinating reading. Though there are many anomalies, this is pictorial journalism and science combined at their best.

There is a reasonable "middle ground" in the controversy between those who view cataclysm as a viable part of nature and those who will relegate it to only the earliest stages of planet building. If we had cataclysmic beginnings, can we not also have cataclysmic middle and ending? When Pangaea was the whole ball of wax, so to speak, what was on the other part of the globe under all that water? Well-lubricated, smooth basalt stone for continent sliding, I'd suppose. It is a well-known fact of paleomagnetic studies that the poles of our planet have "flip-flopped often—north has become south and south has become north at least 170 times in the past 76 million years"; that statement is taken from *National Geographic*. All right, if the poles have "flip-flopped," what force caused it? Additionally, there is evidence that the poles didn't necessarily reverse, they *shifted* from place to place relative to the continental masses. There's really only one answer for such a phenomenon—continental movement. And we don't necessarily mean slowly creeping—we mean sudden, violent movement.

Back to the drawing board. The earth's crust evidently rests upon a layer of molten matter that is apparently heating up. It's sort of like the lid on the poaching pan when water is boiling underneath. The lid jiggles a little to let some steam out, otherwise we'd have a pressure cooker. That's the way I see the earth's crust; a rigid mass, on top of a cross between a pressure cooker and a poaching pan, that is constantly adjusting and readjusting to the pressure surges by volcanic eruption, earthquake, and even drift.

As Hapgood and many others have ably pointed out, the molten magma is viscous or tarlike, and it tends to resist sudden jolts, but it does eventually yield to steady pressure. Most likely it is also very slick in places. The crust, which is uneven, fractured, and rela-

43

tively weak, squeezes back against the magma because of a mysterious force called gravity; thus, we have pressure against pressure. Tectonic plates are outlined by cracks in the crust where most of the pressure is released during normal "stable" conditions. There is also a factor that physicists call "isostasy" which means that whenever the rigid material "jiggles" a little to release the pressure, there will be a period of settling back into the viscous material until all's balanced between crust and magma. This is perhaps the reason so many major earthquakes sustain several severe aftershocks.

Although I'm not a seismologist, I know a little about shock waves from practical experience. If you were to take a length of brittle stone and apply steady, but not sudden, pressure until the stone snapped, you would create a shock wave at the instant of cracking. Should this piece of brittle stone happen to be part of an intact sidewalk, the ants near that crack would really be shook up. The same applies when the crust cracks and our puny little cities shake apart. This is not as oversimplified as it seems, but unfortunately we cannot build a laboratory complete with spinning spheres in a vacuum to test how, *in vitro*, all this really works. If we were able to spin a ball of matter without axles and play with the temperatures of materials inside, we might succeed in solving some perplexing problems.

Hugh Auchincloss Brown, a doughty old electronic engineer, incurred the wrath of the geological establishment by writing a book tilted *Cataclysms of the Earth*. Brown, a freethinker and obviously nobody's fool, developed the theory that the weight of the ice caps could cause an imbalance in the centrifugal forces of the earth's spin and bring about a careening of the entire globe. Although there were a few holes in his thesis, his ideas were solid, and it pays to take a long, hard look at the weight of the ice at the South Pole today.

Unbalanced centrifugal effects are clearly illustrated

44

every day when people wash clothes in spinning washers. If heavy objects get lumped together on one side while the machine is rotating at high speed, you will encounter vibration and shaking, often of sufficient force to make the whole machine wobble or even to pull up floor mountings. Brown stressed that this same kind of imbalance is happening to our planet as it whirls around on its axis at more than 1,000 miles per hour at the equator. He pointed out that the earth wobbles slightly as it spins, causing an approximately 50-foot deviation from the center of the axis at each pole. These "wobble deviations" eventually form a cluster that is completed, according to Brown, every fourteen months. This means that if you were in a space capsule hovering directly above the pole and had a golden screw to mark the precise point of the polar axis, you would see, over a fourteen-month period, a clusterlike movement as illustrated in photo section. Relatively speaking, this deviation is very slight, and nobody, repeat nobody, really knows why a spinning planet would do such a thing. Brown has pointed out that this wobble means the planet is spinning slightly off center most of the time.

In addition to the wobble there is the ice factor. At the South Pole, the ice is not well centered, and the Antarctic ice cap is much, much bigger and heavier than most of us realize. The actual point of rotation, or pole, is more than 300 miles off the estimated center of gravity of the polar mass (see illustration in photo section). As a result this huge mass of ice is rotating eccentrically—much like a rug in a spinning washing machine. This fact is one of the vital keys to this thesis and to our future and, as such, it is no insignificant matter. When the two land masses are compared, Antarctica is nearly as big as all of North America and it is almost entirely covered by an enormous ice sheet. There are moutain ranges on Antarctica which compare in size with the Rockies, yet the ice buries them except for an occasional peak. Polar experts maintain that the *average* depth of the ice sheet is a *mile* below the surface and in many places it appears to reach two miles. Estimates are that there are approximately six million cubic *miles* of ice on that continent, and storms add nearly 400 cubic *miles* to that mass each year.

I realize there have been reports in recent years that the ice is melting or retreating, but the evidence supports growth, in fact, very rapid growth over the past 10,000 years. I remember being told in seventh grade science class that the ice cap was melting, and then seeing in a newsreel at the local theater how one Antarctic expedition a few years earlier had built a radio tower only to have the ice and snow cover it up. I would speculate that the cap is melting around some of the edges while building up in other spots.

The point is there is a single rug in our washing machine and the force of eccentricity is tremendous. However, there is a built-in stablilizing factor, and while everyone is familiar with it, hardly anyone realizes the reason for its presence—the earth's equatorial bulge. We have all learned in school that the earth is not a perfect sphere, but rather more of an oblate spheriod, which means that it's sort of flattened at the tops and puffy in the middle. Obviously, the centrifugal force of the rotating spin acting on the layer of magma is the reason our "rigid" crust has a twenty-six-mile bulge around its middle. As pointed out earlier, this bulge acts like the stabilizing rim of a gyroscope.

You may have heard that our planet is actually a little "pear-shaped." This is accounted for by the uneven weight factors of ice at the two poles. The South Pole is heavily laden with ice, putting tremendous pressure on the magma beneath Antarctica, while the North Pole is covered with ocean and ice floats (except for Greenland), applying much less pressure to the magma at the "top" of the planet.

Brown made another interesting observation when he

46

pointed out that the crustal readjustment following a "careening" of the globe would amount to one-half the distance of the bulge in each hemisphere opposite the equator, or roughly thirteen miles of adjustment room. It just so happens that thirteen miles is very close to the maximum difference between the highest mountain peak and the deepest ocean trench. While it doesn't prove anything, this explanation does help to make sense out of the factors inherent in "isostatic" adjustments within the earth's crust. Those scientists peering at the geology of Mars and Venus via our technological wizardry should consider the spin of those planets as a factor in whether they feature "techtonics" or not.

Hapgood worked in collaboration with a brilliant mathematician and engineer, James Campbell, who computed the mechanics for a crustal displacement based on the weight of the ice and its imbalance. Brown had the right idea, but he computed only for the wobble and not for the 345-mile eccentricity in the bulk of the ice cap. The imbalance created by the wobble alone probably could not overcome the stabilizing factor of the equatorial bulge. Evidently it takes quite a jolt to knock our planet off kilter as we wend our way merrily through the cosmos, and that in itself is comforting.

What all this means to you and me in the coming twenty-three years is a matter for serious discussion later, but keep the centrifugal effects in mind. At this point we are merely establishing the potential physics for crustal movement that could trigger tremendous changes in the life which clings tenaciously to that shaky surface.

Entire species of animal life, along with entire civilizations of man, have been wiped out in the past. If we stop right here and fail to assume that such shifting of the continental plates can occur, we are forever faced with a number of anomalies, of unanswered questions relevant to the facts of the earth sciences. This is precisely what the traditional uniformitarians have done—stopped.

From this point on in this thesis regarding the life and death of planet earth, we *assume* that the weight of ice and other factors, such as collisions with planetoids, can definitely affect the stability of the rigid crust as it sits atop its pressure cooker.

Back, back into time we go—back millions of years if you wish, back to the times in which we can't prove or disprove a thing. Let us assume the time frame is millions of years ago. The planet is about one-fifth land and four-fifths water with all the land lumped together in a big island called Pangaea. One of the poles, let's say the South Pole, is mostly covered by land, while the other pole is entirely submerged. Various forces are at play including the "throwing" effect of centrifugal forces inherent in the huge ice cap at the South Pole; gravitational forces; and pressure from under the crust. There is plenty of volcanic activity as well, just as there is today. And there are special magnetic, electric, cosmic, and mystic forces also at play—just as they are today. We don't know much about them, but what's new?

Biologically we can assume one of two things: either God has instilled the qualities of life into the chemistry of the planet and is helping life forms evolve; or somehow the atoms making up matter on this planet have managed to arrange themselves so that eventually they could have abstract quality and conceive of themselves. One theory is probably wrong, but regardless, life forms are evolving and evidence indicates that evolutionary changes most certainly have occurred. The irregularly scheduled cataclysms for this planet make adaptability and natural selection a must if species are going to continue to keep up with the changes.

Then, as now, times goes by. Mountain ranges are thrust up by the pressure cooker below. Sometimes the crust is "tilted" to make mountains (like the Sierra

48

Nevada range of California), and sometimes volcanic action does the job (like Paricutin in Mexico), but most often it's a combination of both forces. Every now and then a chunk of cosmic real estate bangs into the earth and raises havoc with that "isostatic" quality between crust and magma, perhaps moving the crust a little and helping separate the big island of Pangaea by lowering some crust and raising other parts. You can visualize the happenings by merely thinking of thousands of years, sometimes a million years, as short blocks of time, with the pressure cooker intermittently belching lots of lava and so forth as punctuation. Every once in a while it all comes together: the weight of the ice, alignment of planets in the solar system, or a huge meteor, and whumphf! The crust displaces; the drifting continents suddenly pull apart; volcanic dust fills the atmosphere; tremendous amounts of steam make the humidity unbearably intense as magma meets water. Imagine these conditions. The atmosphere will also experience tremendous changes, thereby affecting weather conditions. Scientists examining ice chunks that have been around for thousands of years in Greenland and Antarctica noted that the proportion of carbon dioxide in the air was far greater during certain periods than during others. Of course it would be greater if the tremendous volcanic activity associated with cataclysms is taken into consideration. Any volcano emits vast amounts of carbon dioxide into the air. When Krakatoa blew its top in 1883, it left its dust and carbon dioxide marks on much of the earth for several years. But as massive as that blow-up was, it was child's play compared to volcanic action that is associated with cataclysm and subsequent isostatic readjustment.

A sudden influx of carbon dioxide naturally changes the proportion of oxygen to carbon dioxide in the atmosphere. This change can be detected in centuries-old ice. Uniformitarians, when confronted with this fact, wrote: "Probably this ice was formed back in the Pleis-

49

tocene time, when cold climates may have curbed the photosynthetic activity of green plants over large parts of the earth, resulting in a slight lowering of the oxygen content of the air." (*Science.* January 20, 1956.)

Again we see the unrelenting struggle by uniformitarian paradigm apostles to explain evidence so that it fits. As we have already noted, there seems to have been a "climatic optimum" following the last ice age. Carbon dioxide gases will remain in the atmosphere for a long period of time, and it is known that sunlight is not affected by heavy concentrations of carbon dioxide. But the radiation of earth's heat back into space is affected, and the evidence of carbon dioxide gases in quantity would, therefore, account for a considerable warming trend. It is likely that these effects accelerated climatic changes, thereby greatly accelerating geological processes (erosion), and perhaps greatly influenced biological processes as well, especially in the case of green plants which, of course, thrive in a carbon dioxide atmosphere.

Then there is the process of mountain building. Campbell and Hapgood went into special detail on the folding and fracturing of the crust that would occur in a displacement. The movement, started by the centrifugal thrust of the weight of ice, would necessarily be north-south since we spin east-west. The crust movement toward the equator would naturally force an expansion, opening up major faults in a north-south direction while concurrently further defining many lesser cracks or faults, like San Andreas in California. in more of an east-west direction. From the equator toward the pole the crust will naturally be forced to contract, causing crustal folding with major faults going in an east-west direction. (See illustrations taken from Hapgood in photo section.) Despite the criticism and general badmouthing received by Hapgood and Campbell for their work, the presentation they made is concise, detailed, and thorough; it certainly holds up to the most severe scientific scrutiny.

That such folding and fissuring has occurred in the earth's past cannot be denied. Campbell even points out that this thesis could be a useful tool in mining, because the fissures would naturally be filled with magma and all the minerals such matter would naturally contain. His point gives us a chance to make an interesting digression. Geologists can locate oil and oil companies can drill for it, but nobody knows for sure how the oil pools came to be located in one spot as opposed to another. With the evidence mounting in favor of cataclysmic shakeups taking place, here's one hypothesis which explains how the earth's crust manufactures oil. When magma comes charging up into a fissure it begins to cool. As it cools silicon-cyanide is one of the very first crystals to solidify; pockets of these practically pure crystals form deep in the crustal rock. Isostatic adjustments will now cause water to penetrate into these crystalline pockets and chemical reactions will occur. As the oxygen in water is taken up, silicon dioxide is formed and the hydrogen (from water) latches onto the carbon present in the cyanide to form hydrocarbons. This hypothesis was formulated by Richard Kieninger, author of a highly recommended book titled *The Ultimate Frontier* (his chemical theory was related to me in a private conversation and is not included in his book). If it is correct, this means oil is constantly being "manufactured" within the earth and therefore isn't necessarily just left over from fossilized ocean life.

Once again we return to the drawing board where we have constructed a developing earth, rotating and orbiting and shaking along toward the present. It is easy to see a number of correlations between the sciences when we consider cataclysmic data. For example, when the molten rock spews up from the pressure cooker and is forced out the top of the crust, the magnetic particles will align with the position of the poles

51

as they cool. This leaves a clearly marked trail, telling us where the poles were relative to the land masses when that hot stuff erupted. It helps solve the great enigmas of biological history such as finding the remains of big lizards in what is now the frigid Arctic. Cold-blooded reptiles can't live under such conditions, yet paleontologists found large Permian reptiles fossilized along the Dvina River in Russia, just below the Arctic Circle. Unexplained changes of climate which are indicated by paleonotological discoveries are readily explained by displacement.

Even traditional geologists would find solutions to strata anomalies by considering crust displacements. Only rarely can deposits be found that indicate long periods of time without interruption, yet there are thousands of cases where "uniform" conditions were obviously interrupted. Some of the best evidence of these interruptions comes from one of the uniformitarians' best arguments against such drastic changes—coal beds. Seams of coal are well known and well documented by the preponderance of literature on the formation of coal. Curiously, a knowledgeable geologist has this to say: "It is worthy of notice that the stratified beds between the coal seams are of marine and not of lacustrine origin. . . . If, for example, there are six coal seams, one above another, this proves that the land must have been at least six times below and six times above sea level." (James Croll, *Climate and Time*, 1875.)

In other words, the geology of coal beds indicates sudden changes in crustal and weather conditions rather than long periods of uniform erosion and so forth.

Now, let us assume it is just over a million years ago, and the evolutionary processes have brought about a creature ideally suited to utilize a number of abstract qualities not found in the make-up of other creatures. This dramatic moment, despite all atheistic protestations, most assuredly could not have come about as a result of chemical reactions alone. The evolutionary processes had creative help—that's a theory that withstands scrutiny and handles a lot of "anomalies." Curiously, we generally accept or believe in the thesis of a God, but our sciences simply overlook such a notion in the construction of paradigms.

Regardless of the emotional arguments that have emanated from many brilliant minds on both sides of the "origin of man" controversy, the fact remains—a thinking animal did come along and did dominate and reshape the environment. By this time the original lump we have seen fit to label "Pangaea" is now unrecognizable. There are a number of continents, including a massive plate covering nearly all the area now comprising the Pacific Ocean. This particular plate includes all of present-day Australia, New Zealand, and parts of present-day North America. Someone, a long time ago, carved the name for that old continent on some rock tablets and the writing has been translated "Mu." The hue and cry from orthodox archaeology is that "Mu" is a "mistranslation" due to the excesses of August Le Plongeon and Brasseur De Bourbourg when they translated some Mayan carvings known as "codices." Since experts are still in the process of translating most of the Mayan carvings, who is to say whether the Frenchmen were wrong?

Another name for Mu is Lemuria and today the two are used interchangeably. Lemuria derives from the Latin root for "ghost," and one story has it that when a small land mass popped up in the Indian Ocean off Ceylon and had apparent ruins of temple columns and other remnants on it, British investigators called it a "ghost town" in Latin. Lemuria was meant to be the reference term for land beneath the Indian Ocean, but somehow it got redirected to the Pacific. Regardless of the name, the general opinion is that such a place is even more fanciful than Atlantis.

Though I can't cite conclusive proof at this point, it makes more sense to accept that Mu and Atlantis existed with civilizations on them than it does to ignore the evidence and figure that civilization is less than 8,000 years old. I've read two authors who wrote about Mu, only one of whom might be considered famous. James Churchward is well known and his books about Mu, written in the late twenties have been reissued in paperback recently. Churchward became "metaphysical" and in so doing really ticked off orthodox scientists. Despite many flaws, some of his research has considerable merit.

The second author is a great deal more "metaphysical." In 1952 an unobtrusive organization published *The Sun Rises*, a book written by their president and founder, Robert D. Stelle. It is the story of the formation of man's first civilization on the continent of Mu about 80,000 years ago. Dr. Stelle's information was gleaned from a mystical source referred to as the "akashic record," and is therefore considered fictional by arrogant experts today. Stelle died shortly after the book was published and never finished his sequel which was to have been titled *The Sun at Zenith*, detailing the civilization of an extensively developed Mu.

Unlike Churchward's *The Lost Continent of Mu* which deals in archaeological evidence and subjective interpretations of certain stone carvings, *The Sun Rises* reads like an action novel and would surely be listed as "science fiction" by most librarians. However, many people, myself included, find that Stelle's story rings true. Not only is it good reading, but it is richly endowed with philosophy and social commentary that, if applied in any way, would serve the world well today. In fact we would do very well if society adopted and practiced the "10 Lemurian Laws" cited by Stelle within the text. A slightly edited version of those 10 laws appear in the illustrations section. There, too, is a reproduction of Dr. Stelle's map of ancient Mu which may be contrasted with Churchward's map. Stelle's map is much more interesting than Churchward's for several reasons, not the least of which is how he came to acquire this information.

At this point an explanation of the "Akashic Record" is necessary. To understand these metaphysical records one must first grasp the concept of "planes of existence," a concept foreign to us in the Western World, but acknowledged by most in the Orient and Middle East.

In our day-to-day life we exist on the "physical" plane and are bound by the laws of physical science. The next plane is called the "vital" or "Etheric" plane by many, and it is the level of existence which acts as the "glue of the universe." The etheric plane, or "ethers," form a matrix for the atoms of the physical plane. It is said that many people can actually preceive the "etheric glow" that surrounds physical objects and beings. The third plane is the "Astral" plane where it is said we dwell between incarnations into physical existence. The fourth plane is the home of human thought and is often called the "Mental" plane. These levels coincide and interrelate, but always adhere to universal rules, or law. Every human thought emanates from the fourth plane and the "image of God" concept is rationalized by this view. Every human thought which precedes every action, is indelibly marked on the "Etheric" level because human thought profoundly affects the "ethers." Thus the concept of "mind over matter." The "Akashic Records" consist of these indelible thoughts in the "ethers" and a person who has learned the proper techniques—usually progressing along with character growth—can comprehend the thoughts, and therefore the deeds of past human activity. One does not actually "read" the Akashic Records, I am told; it is more like clairsentience where the reader attunes to the records like a phonograph needle in the grooves of a record.

It is said that Dr. Stelle obtained his information about ancient Lemuria from the Akashic Records. Many individuals have claimed to read the Akashic, or "Book of God's Remembrances," though few have left works attesting to the veracity of their claims. In addition to *The Sun Rises* by Dr. Steele, there is the *Aquarian Gospel* by LEVI, which is the story of Christ's life and activities during times omitted by the scriptures. Levi H. Dowling, 1844-1911, obtained his information regarding Christ from the "Akashic Records."

Generally speaking, you and I, are stuck with the choice of either taking or doubting the mystic's word for the source. All we can do is read the material and judge its veracity from our own experiences, or learn to do the same thing—but then would anyone believe us?

The Akashic Records have been blamed for a great deal of verbiage, and this is unfortunate because it has demolished their credibility in a world in which credibility is far more important than truth. Nevertheless, Dr. Stelle somehow obtained information regarding the tectonic plate beneath the Pacific; information that is theoretically more accurate than outlines made by seismologists today; information that he could not have obtained from any geodesic or oceanographic survey since none had been completed when Stelle drew the map!

Churchward's information, on the other hand, was gleaned in an unusual, but not metaphysical way. As a young military officer serving in India, Churchward was introduced to some stone tablets of exceptionally ancient origin. He was told by a Hindu scholar that they were Naacal, probably originally from Burma, and that these were the writings of colonists from man's earliest civilization which developed on Mu.

Churchward spent not five or ten years, but fifty years trying to prove archaeologically that the Naacal

56

tablets indeed proved the existence of a civilization on Mu and that a cataclysm (he calls them "magnetic cataclysms") indeed destroyed that continent in the Pacific. During his search he met an archaeologist named William Niven who uncovered the ruins of ancient cultures at a site about twenty-three miles north of Mexico City. Niven is not generally credible among archaeologists and, like Churchward and von Daniken and others seeking to prove a theory, he may have made erroneous assumptions about his discoveries. For example, both Niven and Churchward assumed certain artifacts had to have come from Mu or Asia because they were made of jade; they were operating under the erroneous assumption that there was no green jade in Mexico. Since their time, however, many jade artifacts have been found in Mayan digs.

Whenever an "orthodox" scientist makes a faulty assumption or two he is forgiven, but whenever a "kook" makes one, the baby is tossed out with the bathwater so to speak—meaning everything Niven and Churchward said was tossed out because of their inaccuracies. Niven did uncover some interesting facts, and Churchward did an excellent job correlating some facts even though he was working primarily with the remnants of Mu's colonies rather than Mu itself. That's like describing the United States with artifacts dug up in the Virgin Islands!

Churchward's map of Mu suggests a much smaller continent than Dr. Stelle's map and the "ring of fire" suggest. Looking closely at Stelle's map, we see that the outline of Mu as he claims it once stood above the ocean's level takes in all of Australia, Tasmania, and New Zealand on the farthest southwestern section; it compasses Easter Island at the extreme southeast, but interestingly does not take in the teeming Galapagos Islands, so dear to Darwin's evolutionary research. It includes all of the western United States and Canada at its extreme northeastern reach, then runs right down the Aleutian Islands and along the now well-known Pacific "ring of fire" to include New Guinea, but not the rest of the East Indies.

Today, if you follow the volcanic ring, you find a discrepancy in Dr. Stelle's map. You will see tremendous activity much closer to Japan than Stelle's information takes the Muvian border. You will see double volcanic activity between China and the ancient continent and a tremendous burst of activity across New Guinea and eastward almost to mid-ocean, then downward to New Zealand, thereby appearing to exclude Australia and Tasmania from that ancient plate. Also, today's seismologists assume that the San Andreas fault is the line between the North American plate and the North Pacific plate, whereas Dr. Stelle draws that border well into Montana and Wyoming.

I'll accept Stelle's version, not because I'm argumentative, but because his outline of Mu fits *all* the available evidence. For example, there are only two areas of the world with soils and certain flora that compare at all favorably with soils and plants found on Australia and New Zealand and New Guinea—those named and parts of California. So, I contend, Stelle was correct to include Australia in his map. Evidently, something cracked the original Muvian plate to account for all the seismic and volcanic activity in the southeastern Pacific today. Certainly the stresses of a submerging continent could cause such a rift. After all, Pangaea broke up, didn't it?

Evidence for Dr. Stelle being more accurate than seismologists with regard to the area of collision with North America may also be contested, but it is there. The seismologists' version states that about 30 million years ago the Pacific plate "drifted" into the North American plate in the vicinity of Los Angeles. The impact thrust up the Sierra Nevada Mountain Range, and the "trenching" factor accompanied by continual upward drift by the Pacific plate accounts for the coastal range. This concept, and again I am stressing that it is merely a concept put forth by specialists, is feasible over millions of years, providing there were some genuine jolts to account for the forces of mountain building. Since I intend to show that Mu sank suddenly and violently only about 30,000 years ago, I must consider the presently accepted hypothesis as pure bull.

When that particular displacement occurred about 30,000 years ago, the direction was apparently predominantly northward at a more rapid pace for Mu than for North America—later we shall see the mechanics of this and how parts of the globe most certainly move faster and farther than others. The collision between Mu and North America was terrific; all kinds of magma belched up and spilled all over most of the area now known as British Columbia, Washington, Idaho, Oregon, and northern California. The Rockies and the Sierras could well have been forged by the impact. One suggestion in support of this theory lies in the way the Sierras are steep on their eastern slopes, indicating the impact that forced them up came from the east. Naturally the force of the collision cracked the plate in several places and one of the cracks, which clearly shows the direction the continent was moving when it bumped into North America, is very tiny, but very active due to pressures brought about by contraction. (Remember Campbell's work showing contraction in the section moving away from the equator?)

For every action there is a reaction that is equal in force and opposite in direction—you've heard that law expressed hundreds of times before. Well, if two continental plates smack together, it doesn't make physical sense that mountains will go up on one plate and not be thrust up on the other. Of course such an anomaly did *not* occur. Today in Montana, Wyoming, and Idaho, and parts of Colorado and Utah, oil explorers work in a geographical vicinity known as the "overthrust." That overthrust may be excellent evidence for two things: the collision as I have described it and the production of oil as Kieninger described it. Parts of the continental plate bottoms are naturally chewed up and broken off and squeezed down into the magma to become molten "geochemistry." I'm sure that a specialist could make a far more accurate, far more detailed report of the collision between the two continents than I have, but it would probably be far drier and certainly no more realistic.

In *The Sun Rises* Dr. Stelle does not make any attempt to prove his statements about the size and shape of Mu. Neither does he try to offer the evidence that the inhabitants of the first civilization were excellent stonemasons, or that for more than 50,000 years most of the buildings, roads, and canals were built of heavy stone. Though I intend to argue that the Lemurian civilization was the greatest yet advanced on earth, they did not build massive steel and concrete skyscrapers. However, since when have such monstrosities of technology been indicative of an advanced stage of development? Followers of Stelle's teachings point out that the art of masonry was brought to its utmost precision by the Lemurians—a precision not matched even today, but hinted at by the builders of the Great Pyramid of Giza in Egypt. The early stonework of the Lemurians was roughhewn, but built to last, and today remnants of bassalt block edifices, temples roadways, and canals can be found all over the Pacific Islands as testimony to the facts.

Churchward summed it up nicely: "Throughout the length and breadth of the Pacific Ocean are scattered groups of small islands. On scores of them are the remains of a great civilization. There are great stone temples, cyclopean stone walls, stone-lined canals, stone-paved roads, and immense monoliths and statuary—works that required continental resources and workmen of skill. Yet, we now find them not on a great continent, but spread on mere specks of land inhabited by savages and semisavages."

On the island of Ponape in the South Pacific, remnants of an ancient stone-building culture attract tourists today. The ruins are called Nan Madol, and guides, reciting their spiel by rote, estimate the buildings back perhaps 1,100 years. This is anthropological drivel at its worst; basalt stone is among the heaviest building material in existence, and the peoples of Oceania have had only wooden tools to work with for the past 10,-000 years. How, then, could they have quarried, carved, and fitted such blocks into place? As Churchward stressed, it certainly is not merely an isolated anomaly. Rather, basalt ruins can be found on just about every island that exists within the boundaries of Dr. Stelle's map of ancient Mu. Curiously there are no similar cyclopic ruins found in the Philippines, an island area noted for ancient cultures but lacking examples of this particular type of stonework.

Barren Malden Island, located practically in the center of ancient Mu, has basalt roadways leading off into the Pacific in several directions. Now, I ask you—could those ancient road builders have deliberately constructed pavement into the ocean? If so, it was one helluva big job because the same roadway turns up again on Raratonga about 1,500 miles away. And similar stone ruins can be found in the Marshalls, the Gilberts, Tinan and and Guam, Samoa, Fiji, Hawaii, Tonga, Tahiti, the Marquessas, and, of course, famed Easter Island. As you can see by the maps, Easter Island with its fantastic stone head carvings was a distant outpost on Mu. The area would have been at least 6,000 miles from the civilization's capital, Hamukulia, located near the present Hawaiian chain. That last part is, of course, pure conjecture, but the evidence presented on Easter is solid fact.

While the massive stone faces of Easter Island have captured the most attention, it is the hieroglyphic writings on stone tablets found there that really provide a mystery. One expert feels the writing matches glyph writing found in the Americas to some extent; another says certain characters match with "Maori" glyphs from New Zealand; and Churchward liked the translations by a man named W. J. Thomson who supposedly had the aid of a "very old native who understood the symbols." This is typical of the kind of "excesses" that turned scientists off Churchward. Evidently some of the natives did have an idea of what "their own" ancestors claimed about the writings, but none of them could convince language experts they really knew the full meaning of the ancient glyphs. A solution to the mystery was found in concentrating on the massive stone heads while, for the most part, ignoring the tablets. To my untrained eye, however, the glyphs look like a hodgepodge of Arabic, Japanese, Cuneiform, Egyptian. . . .

What we are going to do at this point is simply accept the idea that the evidence seems to point to a continent existing in the Pacific a long time ago, and we are therefore incorporating such a notion as part of our new paradigm. The datings which confirm that the Wisconsin ice age began about 30,000 years ago also suggest that this continent, Mu, vanished in a violent readjustment of the land mass at that time, leaving only traces, colonies, and, of course, survivors.

CHAPTER 4

The Invincibility of Culture:
The "Hard" Evidence for
Atlantis and Mu

When a culture is alive and well, it has a habit of spreading its traits like tentacles to influence neighboring cultures. Just look at the way our culture is spreading asphalt, for example. Cataclysms, however, have a way of wiping out flimsy things like cultural traits so that descendants of the survivors can have honest doubts about their predecessors ever having existed. It's easy to imagine how an ocean pouring over a continent can wash away evidence of civilization; just spray an ordinary garden hose at a table setting and you begin to get an idea of the destructive force of water. Now add volcanic action and horrendous earthquakes and your conclusion will be incontrovertible. It is impossible to "prove" that a civilized culture existed beyond our present paradigm.

Fortunately for this thesis, one of the cultural traits exhibited by the Lemurians was stone building—stone building of super-colossal size and strength. We find evidence of their handicraft, not only clinging to tiny island dots in the Pacific, but also along the west coast of South America, which the Lemurians evidently colonized.

Archaeological experts are at a tremendous disadvantage when they are confronted by the cyclopean ruins found in Peru and Bolivia because they are stuck with pieces to a puzzle that don't fit the form. The ruins of Peru and Bolivia have been called "impossible" by many investigators who looked directly at their obvious possibility. There is no reasonable explanation for the gigantic ruins within the accepted paradigm, so the assumption that the Lemurians colonized and built in this area does have something going for it.

The technique of stacking huge stones atop one another so that a structure was held together by sheer weight and the shape of the building blocks is found throughout the Pacific ruins and also among the ruins in South America. At Sacsahuaman, Peru, there is a wall of gigantic stones where some of the blocks weigh (individually) up to hundreds of tons. These huge blocks are patchworked atop one another with amazing stability, and it is mind-boggling to imagine how mere humans could handle such stones without cranes, trains, and automotive power.

Another site of megalithic construction is found at Ollantaytampo, Peru. Here the masons from Mu managed to build a wall of blocks, each weighing upward of 150 tons, that fit together with a precision similar to that achieved by the builders of the Great Pyramid in Egypt. Today the wall is found atop a 1,500-foot cliff; if the cliff existed when the wall was built, how did workmen manage to transport building blocks of that proportion up to the site?

Finally, there is the famed site at Tiahuanaco on the shores of Lake Titicaca in Bolivia. First of all, this ancient city, a megalithic marvel in itself, is located some 13,000 feet above sea level in the Andes Mountains.

The height is dizzying to visitors without oxygen masks, corn won't grow at that altitude, and even cats shun the place, indicating that the environment is not the original environment of the ancient dwellers of this place. It wasn't! Researchers have shown that Tiahuanaco was once a seaport, so evidently something pretty violent came along and thrust the mountains up.

Like most ancient ruins, there has been a certain amount of pillage connected with Tiahuanaco which eliminated some of the evidence of highly advanced civilization. It has been shown that the Spanish conquerors took silver and other metal work associated with the stone construction, so it is impossible to investigate what certainly must have been a high level of metallurgical knowledge. And, just as the blocks of the Great Pyramid of Giza were looted by Moslem builders in Cairo, the ruins of Tiahuanaco became a quarry for builders in La Paz, Lima, and other towns in the vicinity.

Since the Lemurian cataclysm apparently occurred about 30,000 years ago, give or take a few thousand years, my estimate of the age of Tiahuanaco and other "pre-Incan" ruins is at least that long ago. There is no atomic dating system for stone, and carbon dating the bones of extinct animals found in the vicinity doesn't necessarily correlate with the civilization's era, so it's a matter of guesswork. Today's estimates range from a ridiculous 1,500 years ago to more than 15,000 years ago—and these guessers are walled inside the paradigm. Some pottery fragments, however, bore drawings of extinct animals, hinting at the antiquity of the sites.

The archaeology of these ruins is so fantastic that experts keep digging themselves into a deeper hole of mystery, and in so doing, they strain our present historic paradigm. The so-called "Gate of the Sun" found at Tiahuanaco has become the focal point for some far-out theories, especially the notion that our forefathers were astronauts from outer space, a thesis put forth eloquently, if not entirely accurately, by irrascible Erich von Daniken. He envisions evidence of space age technology in the carved anthropomorphic figures. Instead of merely birds and serpents and so forth, von Daniken and his followers see in the carvings electronic gear and astronaut garb. This view is not so farfetched as orthodoxy would have us believe. The idea of looking at ancient drawings and carvings in a different light is certainly a worthy one, and if anything, the reevaluation of the messages in the drawings may clearly indicate greater knowledge of certain technologies in the past than we ever imagined. I do not accept the idea of the outer space connection, and later, when we discuss the purpose of it all, you'll understand why I take this position.

Anyway, getting back to the diffusion of cultural traits as a way to determine whether vast civilizations truly existed in "prehistoric" times, we find that tremendous stone masonry is also found in the "cradle" of civilization. Even though it is located in the midst of the most dug-up, archaeological region in the world—the Middle East—the ancient stonework found at Baalbek, Lebanon, is equally as mysterious as the South American ruins.

Then there is Stonehenge, the circle of huge stone monoliths in England that has intrinsic astronomical significance; unbelievable, battleship-size stone carvings in Ethiopia; and the thoroughly fascinating wonder we call the Great Pyramid, which in itself is more than enough for a complete book.

The stone-building trait most likely carried over to "lost" Atlantis, and some of the masonry now being discovered in the ocean waters off Bimini, Bermuda, and other western Atlantic isles may soon prove that assumption.

When it comes to the diffusion of cultural traits, two elements have combined to make it more difficult to trace a Lemurian culture than it is to trace an Atlantis

66

culture: time and a much bigger, more violent cataclysm. My reasoning is as follows: The Mu cataclysm ushered in the North American ice age and was responsible for thrusting up new mountain ranges on North and South America, so it obviously occurred much earlier and was far more violent than the phenomenon that caused the sinking of Atlantis.

Because Atlantis is much closer to our time period, relatively speaking, and because it did not sink so violently, there is a vast amount of evidence available to indicate that such a place not only existed, but had such an advanced civilization that later cultures on both sides of the Atlantic Ocean were influenced.

So much has already been written about Atlantis that it is difficult, if not impossible, to come up with something new to say. However, I think that you'll find that much of the following is new information.

Every discussion of Atlantis traditionally begins with the writings of Plato, the famed Greek scholar and philosopher. Since I'm certainly not one to go against tradition, I'll also begin with Plato's famed descriptions of the "legendary" Atlantis. Plato, we have all been taught, was Aristotle's teacher and a pupil of Socrates; he therefore ranks among the key figures in the study of the Golden Age of Athens and its effects on our world today. As usual, our concepts of what went on before our times is distorted. Those Golden Greeks are credited with having given the world certain mathematical and scientific concepts, as well as astronomical assumptions alleged to have been well ahead of the rest of the world at that time. No doubt about it! The early Greeks were sharp, but historians generally overlook the fact that nearly every one of the brilliant Greeks journeyed to Egypt to learn what the Egyptian priesthood knew before they came home to Olympus and shouted "Eureka!" This is not meant to demean the importance of those great Greek thinkers; it is merely my way of illustrating how even our historic assumptions can get distorted. Most of the astronomical facts about the earth orbiting the sun and being a sphere that rotated on its axis were known to the "secret societies" among the ancient Egyptian priesthood. Most of the cosmology of both Pythagoras and Plato was taken from Egyptian sources. The key difference is that while the Greeks shot their mouths off about the knowledge, the Egyptians kept it locked up for priestly use only.

Quite naturally, then, it should come as no surprise to learn that all the famous remarks about Atlantis credited to Plato's dialogues of *Timaeus* and *Critias* came directly from Egyptian sources. Plato's technique for getting his viewpoints across in writing was to use conversational scenarios between real people. In this way he made the reading come alive and kept his audience interested. His technique does not mean that Plato was "kidding" when he wrote about Atlantis, which appears to be the consenus of today's orthodox opinion.

Aristotle, who evidently became disenchanted with his teacher, Plato, probably did more to screw up our present historic paradigm than anyone when he insinuated that the Atlantis story was pure fiction. Later, the church turned many of Aristotle's dictums into dogma, and Atlantis was shunted to a position outside history's framework. It is possible that Plato was using some "legend" to make his views on history and government more colorful, but he seemed to me to be more likely to want credit for inventing colorful ideas than to credit others.

Regardless, Plato's descriptions of an ancient civilization in the Atlantic Ocean, beyond the Pillars of Hercules (Gibraltar), smack of truth insofar as they are bolstered by cultural similarities found in later civilizations. For example, Plato said there were ten kings of Atlantis and they picked one among themselves to rule over all. A parallel to this novel idea is found in Mayan tradition and also among the Canary Islanders.

Plato was liberal with representations of numbers and sizes and other details. In one instance he elaborates on the requirements for the military each feudal leader of Atlantis was to furnish. In that account he says: ". . . and also, two horses with riders upon them." This indicates domestication of the horse long before our historians have acknowledged it.

He mentions hot and cold water running right out of the earth and hence he correctly described a phenomenon now very much a part of life in Iceland. His descriptions of the gold in Atlantis later inspired the Spanish Conquistadores in their relentless search for El Dorado. The booty taken from the halls of Montezuma is certainly evidence that great wealth could have existed—especially since the Aztecs and Mayans indicated to the Spanish that their ancestors indeed came out of the Atlantic.

Curiously, when Plato mentions the colors of the stones used in masonry by the Atlanteans—white, red, and black—he describes exactly stone still existing on the Canaries. His description of the city plan and port facilities was detailed, and in one instance he says: ". . . ring upon ring of alternating land and water." Thus, he laid out the description of the canal system surrounding the citadel on a hill. City planning can certainly be considered a cultural trait, and evidence of the diffusion of such a trait from Atlantis to both sides of the Atlantic was found when the designs for both ancient Carthage and Tenochtitlan, near Mexico City, were unearthed. The massive stone used in construction is strongly suggestive of a tradition inherited from the even more ancient Lemurians.

Poseidon, the Greek god of the seas, was the number one deity of Atlantis. Now I dispute whether he was an anthropomorphic god or actually an emperor revered by the later Atlanteans. *The Ultimate Frontier,* the book in which I take so much stock, claims that Poseidon was actually Melchizedek, an archangelic entity. We shall explore this interesting possibility in later chapters. The reason it is brought up here is to show how the cult of Poseidon spread to different and "unrelated" parts of the world from a central point that had to be Atlantis.

Along the Gold Coast of Africa near the mouth of the river Niger, there was found a bronze statue of the god Olokun. The natives of Jorubaland who worshipped this deity are descendants of an ancient African kingdom called Ufa, which covered the entire Gold Coast in the past. Many anthropologists suggested that the marvelous prehistoric monuments uncovered in this region smack of colonization by Atlanteans.

Evidence is everywhere and the confusion perhaps stems from the occurrence of other tectonic readjustments which wiped out intermediate cultures. *The Ultimate Frontier* claims that after Lemuria and Atlantis, the next great civilizations of man were Osiris, a valley kingdom now beneath the Mediterranean Sea which coexisted with Atlantis, and Rama, a civilization begun by colonists from Lemuria which existed in India. There is certainly evidence to support such a possibility around the rim of the Mediterranean and also throughout India. And an archaeologist, Hans Fischer, claimed there are traces of Atlantean influence even in China!

So much for idle conjecture; let's get down to the nitty-gritty of cultural diffusion. It can be shown that only one of two possibilities exist to explain the evidence. The similarity of cultural traits in ancient and diverse lands can be explained most rationally by the assumption of Atlantis. It is possible, though highly unlikely, that the Egyptian civilization is much older than we assume and that the earliest Egyptians colonized much of Africa and South America, leaving their ideas behind, but in a slightly altered form. As you can see, that's stretching a point a bit too far, Thor Heyerdahl and his raft not withstanding.

Of all the cultural traits to spread from the central

Atlantic and influence later cultures, pyramid building is the most outstanding. When one thinks of pyramids, one thinks first of Egypt, and the superb edifices located on the plateau at Giza near Cairo, then of the magnificent Mexican pyramids, the pyramid of the Sun and Moon near Mexico City, the pyramids of Chichen Itza and Cholula. The similarities are amazing; in fact these examples make the present-day paradigm ludicrous. We are taught there is no connection between the two cultures (the Egyptian and the ancient Mayans), unless of course you happen to be a Mormon. According to their philosophy, Christ and other "latter day Saints" visited the western hemisphere and influenced the Amerind cultures.

It seems apparent that the pyramid-building trait derived from the "Sacred Hill" of Atlantis. Archaeologists are still uncovering man-made mountains in South America that were obviously built as "sacred hills." It stands to reason that a "sacred hill" can accomplish several things, such as keeping track of the sun's annual high and low points, indicating respect for a higher intelligence or authority (much like church spires) and they might also serve as monuments or memorials for the greatest of the living when they die.

In *The Great Pyramid—Man's Monument to Man*, I stressed how the massive Pyramid of Cheops (a misnomer) is indeed the absolute height of such a cultural trait. The *fact* of such a structure tells us there is much more to this story of man's civilizations and development than mere archaeology will ever uncover.

Of all the books written about Atlantis, my two favorites are *Atlantis—Mother of Empires* by Robert Stacy Judd, a Los Angeles architect who became enamored of the Mayan "motif" and then set out to prove that such a brilliant culture most certainly had to stem from Atlantis; and *The Shadow of Atlantis* by Col. Anton Braghine, a French anthropologist who saw Atlantis and cultural diffusion as the only answer to the

71

many archaeological mysteries being ceaselessly unearthed. Both books came out in 1939 and were compiled during the thirties after years of exhaustive research and personal investigation.

Although Judd may have put too much emphasis on LePlongeon's "translations" of the Mayan codices, no other author has so brilliantly outlined the diffusion of architecture from Atlantis to both sides of the Atlantic.

Braghine's work is far more detailed and anthropological than Judd's and being a Frenchman, he makes liberal use of texts which have not been printed in English. I was especially impressed by Braghine's reportage of the writings of Dr. Luis T. Ojeda of Valparaiso, Chile. Evidently, Dr. Ojeda undertook a detailed study of the "pre-Mediterranean" cultures back in the twenties but unfortunately this scholar's work has not been translated into English.

Among the illustrations in the photo section is a photograph of a bas relief frieze found in the Yucatan by a German archaeologist named Teobart Mahler. (Mahler's photograph was also printed in Judd's book and has recently emerged in a magazine or two.) This stone monument details a long history of the people, according to Judd, and clearly shows a stone structure reminiscent of Chichen Itza, the pyramid with stairs up the middle, crumbling as if shaken by earthquake, and a volcano blowing its top. Imagine the force of an earthquake that could knock the insides out of a pyramid! The escape of some and the demise of others are also depicted. The artist even included a fish to make sure we understand that it's water. One of the little clouds could be interpreted as a flying vehicle if one lets his imagination run free. I'm told, however, that the little UFO-like object is actually a stylized rain cloud, though how anyone *know*s that as fact is beyond me.

Language characteristics linking the two worlds over the dead body of Atlantis are numerous—and, in my opinion, sometimes shaky. It may be a reliable branch of anthropology, but linguistic comparisons don't strike me as the best evidence, yet Atlantis freaks have tried to make a solid case with language comparisons. I'm reminded of the story Herodotus, the Greek historian, told about the Egyptian pharaoh who wanted to prove which group of people were the oldest among all mankind, the Egyptians or the Phrygians. This Egyptian king ordered a baby to be taken immediately from its mother following birth, and placed him among animals so that he could not hear a human sound. The idea was to have the Egyptian C.I.A. wiretap the shed the child was locked in and perceive the first word he uttered. If the child, never hearing a human sound, uttered an Egyptian word, it would mean the Egyptians came first; if a Phyrigian word, it would mean the Egyptians lost the bet. According to the story, the Egyptians lost. Anyway, language comparisons don't really excite me as evidence for Atlantis, except in certain cases or in conjunction with all the other evidence. Perhaps my reluctance to put stock in linguistic diffusion comes from the fact that so many people "play" with words when trying to make a case for some unprovable point.

Braghine, in my opinion, makes the best language comparisons. He points out that Ojeda and other students of the Mediterranean cultures stress that the people we all recognize as the Phoenicians were actually part of a group of city-states dating well back into antiquity, calling themselves the "Carian Union." The union of small kingdoms included Caria, Ionia, Phoenicia, Crete, and Troy and were jointly governed by an illustrious Chaldean character evidently named "Kar." According to Diodorus and other historians of earlier times, Kar was the first law-giver of his people and also an early monotheist. He founded the City of Halicarnassos, which is interpreted to mean "Holy Garden of

Kar," the name "Kar" meaning either "God" or "sacred" or both.

Of course this character Kar is a "mythical" ruler, origin unknown, who sent the Carian influence everywhere. Mediterranean scholars rank Kar with Zoroaster of Persia and Manou of Babylon as a religious reformer-teacher and point out he evidently preceded Moses by a few centuries. According to tradition he became a Phoenician deity, Mel-Car, but nobody knows when he really existed.

I find the connections with Kar to be good evidence for an Osiris civilization in the Mediterranean as well as for Atlantean diffusion.

This legendary figure sent missionaries (cariatids) and workmen everywhere. His name became synonymous with industry or work, as well as with "sacred." The Carian ships were called "carpassios" and were huge by comparison with ancient standards. The root word for ships carrying merchandise reached us, with the usual linguistic modifications through time and place, as follows: *karbas, barkas, barque, carabella, korabl* (Russian). Even the "ships of the desert" called "caravans" have the flavor of this root.

The prefix "kar" meaning "sacred" is much more widespread and is found in the following "sacred places," Carthage, Karnak, Carnak, Khartoum, Carpassos, Carpathian, and Carmel.

To add a little frosting to this word game, we are told that Kar introduced a cult for the "Lord of the Universe" and called this god "Pan" from the cabbalistic tradition. Of course Pan became part of the later Greek and Roman pantheons and one German professor has shown that in the languages of the Pelasgi (original Greeks), Phoenicians, and Carians, the term "Tu-Pan" was used to signify "the Divine Pan." Priests of this ancient cult were evidently called "Sume" or "Sumers." And the plot thickens.

Either the ancient Carians and Carthaginians voyaged regularly to the Caribbean and South America, or the ancestors of both came from a similar cult in Atlantis because we find these remarkable similarities: Caribs—sacred white people; Caribbean—sacred waters; Caracas—sacred city; and all these tribal names which naturally derived from something sacred according to tribal traditions: Cara, Carara, Carou, Cari, Cariri, Carai, Caraiba, Cario, Cariboca, Carioca (also abode of god in Greek), Caralasca, and so forth, ad infinitum.

That the Carians colonized parts of South America, and yet did not evidently influence the Mayans, Incas, or others, is readily apparent. Scholars deciphering petroglyphs (rock writings) in Ecuador have determined that the area around Quito was once a Carian colony. There is a great deal of evidence that the Chibchas, a highly cultured tribal group found in Colombia and Venezuela, were actually Carians.

All the Indian tribes using the prefix "Car" in their tribal names call "White men" by the term "Cari," even though the word for white coloring in their family of language (called Tupe) is "tinga." A really interesting blow to Amerind lore is the idea that the feathered headdresses may have been brought to the Western Hemisphere by either Europeans or Atlanteans. Diodorus tells us that the Carians wore head adornment made of bird feathers.

Legends of the Caribbean natives also tell us that they came to their South American habitats following a "terrible catastrophe" thousands of years earlier. The legends don't indicate where the land was located prior to the catastrophe, however, so Atlantis buffs make unfair claims in this regard.

Among the proofs of Carian linkage is a name change their legend describes. The priests, who are called "sume," told them they were to use the name "Tupi" because it signified the "sons of Tupan," greatest of all deities—and we're right back where we started.

Word games cannot prove Atlantis, but they are certainly interesting.

To me it's apparent that commerce and colonization took place between the Mediterranean cultures and South America as far back as 4,000 B.C. Our history books don't reflect that likelihood any more than they reflect Atlantis. To me it is indicative that the earliest of our "known" seafaring peoples were not afraid of falling off the end of the world as we are told the early Spanish and Portuguese seamen were. If not, why not? Could it be that they knew from their ancestors, the Atlanteans, much more about the earth than we can understand with our distorted view of their ancient world?

Another line of study has been brought in by Atlantis boosters that provides more thought-provoking questions, even though it can hardly be called conclusive evidence that a land mass existed in the Atlantic 10,000 years ago. I'm referring to the curious habits of some animal life. For example, some people feel that the "crazy" behavior of the lemmings of northern Europe, who will occasionally drown themselves by the millions, is induced by an instinctual urge to return to land that is now submerged beneath the sea. It is interesting conjecture, but I don't think instinct can be blamed for such a mass slaughter; instinct usually serves the preservation of a species and is obviously governed by some higher intelligence.

Much stronger evidence for Atlantis is exemplified by the strange behavior of the European brown eels. The following report is taken directly from Bragine:

Now I would like to draw the reader's attention to the very curious phenomenon of the multiplication of eels, which clearly proves that some millenniums ago there was in existence a continent between Europe and America. Aristotle was the first naturalist interested in the multiplication of eels, but, not being able to discover anywhere the spawn of eels, he passed on this problem to the following generations of scientists, who, during the next 2,000 years, did nothing important toward its solution. They could only establish that the eel, a fresh-water fish, yearly leaves its abode in the European rivers and goes in millions to the sea. If their river does not flow to the sea, the eels wriggle across land to another river which does. Once in the water, the shoals of eels disappear somewhere in the ocean until they return to their rivers some months later. It was understood long ago that they go into the sea for spawning, but how and where they performed this process remained a mystery, particularly since only the adults were observed during those migrations and no one had ever seen their young.

The Danish scientist Dr. T. Schmidt especially studied the lives of eels and gave in 1922 the solution of the problem first advanced twenty-two centuries ago. It turned out that only female eels live in the European rivers and their residence there is limited to two years. During this period eels often change their abodes, crawling from one river, or pond, into another. After two years the female eels begin to swim toward the mouths of the rivers, where the males are already waiting for them. Then the separate shoals of eels merge into a large one and start their long journey westward across the ocean.

They swim at a very great depth, and making eighteen miles a day, after 140 days of swimming across the Atlantic, reach the so-called Sargasso Sea, near the Bermudas. This migratory movement is comparatively easy to observe, because it is closely followed by dolphins, various rapacious fish and flights of seagulls. Reaching the Sargasso Sea, the eels disappear in its submarine forests.

It is necessary to remind the reader that this sea is six times larger than the territory of European France, and is covered by a thick tangle of seaweed, which often impedes the navigation of small ships. Our reader will, perhaps, remember what was said about the Sargasso Sea by the cunning Phonecian (sic) seafarers to the pharaoh Nechao: the same conditions exist in the Sargasso Sea today. [Since most of us probably aren't familiar with what the Phoenicians said to the boss, I'll explain: Nechao sent the seafarers out to explore and they came back saying the sea beyond Gibraltar was "too slimy" for sailing. This was probably a cop-out, but it worked.]

The spawning of eels takes place in the Sargasso Sea at a depth of about 1,000 feet, and after this the females die. The newborn eels soon start their trip to the European shores, forming one enormous shoal 260 feet wide and 70 feet deep. At the beginning, the tiny eels are transparent, but at the end of their journey, which lasts about three years, they become green, and finally brown. Then, near the mouths of the European rivers, the shoal of grown-up eels divides in two: the males remain in the sea and the females enter the rivers and begin their two-year sojourn in Europe. Thus the eel represents really two species: the male, a sea-fish, and the female, which spends one-half of its life in fresh water.

The phenomenon of the multiplication of eels is a very important fact for our subject. It proves that at some time a continent with a great river existed between Europe and the Bermudas. Perhaps the Sargasso Sea is the survival of the swampy delta of this giant prehistoric river, and the eels, accustomed during millions of years to spawn there in safety from their enemies, have preserved this habit until today, when not only that river, but the continent itself has disappeared.

Braghine's conclusion is cute. The eel migrations certainly do not prove any such thing, but their actions are food for thought, and such an assumption is probably as good as anything the orthodox sector has come up with to date.

There is just one more area regarding diffusion from Atlantis that I'd like to explore before getting on with the more metaphysical conjecture that lies ahead.

Many "Atlantologists" find a positive link to the ancient civilization in the curious world-wide distribution of certain plants. The reasoning is this: All the plants considered were believed to be found in a domesticated state only, without any trace of a "wild" or naturally evolved state. The assumption was that they were cultivated on Atlantis (or Mu) and the wild state became extinct after the cataclysm(s). The plants in this category are tobacco, wheat, maize, the banana and pineapple.

Von Daniken and the Ancient Astronaut Society members have every right to become excited over prospects for the banana. After all, Hindu legend tells us that wheat and the banana came to earth from "another celestial body" and were brought here by "Manu," or highly evolved beings who are the protectors of mankind. In his book, *The Sun Rises*, Dr. Stelle puts the Manu and this legend into an interesting perspective that we will discuss later.

The reason such a legend exists can be debated, but the origin of wheat and cultivated bananas of the varieties *musa paradisiaca* and *musa sapientur* provide mysteries for botanical science. The banana bush in its wild state is pretty useless, and it doesn't have any seeds for reproductive purposes—it just keeps offshooting to survive. (This is not to be confused with some kinds of similar plants that do have seeds.) Botanists, like other scientists, can be arrogant and I'm sure someone will argue that today we do know the origins of those plants. I'm willing to bet that if such a claim is made, a thorough analysis will show that they only have theories. Even if we were to find a wild species of these precious plants, such as the Russians found in the Himalaya Mountains years ago when they located some *triticum* in a wild environment, it could not prove that those wild species did not themselves derive from cultivated gifts from the "gods." And is it not feasible to assume that a tall, dark, and handsome stranger landing

79

on your primitive shores and showing you how to grow bananas would be considered a god according to local standards? Even if that stranger was from Mu or Atlantis?

The Egyptians believed that wheat was brought to them by Osiris, one of their Manu, and there are many other such legends around the world.

Tobacco, the experts tell me, doesn't belong in that group because the original wild state has been authenticated. We'll grant this one without argument, but . . .

The pineapple is a doozy. A wild species of the pineapple does indeed grow in the Brazilian jungle, but the cultivated variety is found in Asia going back a long time in history. Images of the pineapple are found on Assyrian and Babylonian monuments, for example, However, these images from long ago represent the Western Hemisphere variety of the plant!

It can be said that by using the systems approach and making all the evidence from all the sciences fit a logical pattern, the only conclusion that can be drawn is one that accepts the continents in the Pacific and Atlantic Ocean locations.

There are several "kook groups" trying to locate parts of Atlantis beneath the Atlantic today, and some evidence of man-made walls and such has been spotted under water. But, even if Atlantis is found, we will only have mere remnants of the whole picture.

CHAPTER 5

Creation and Evolution:
From the Garden of Eden
to the Committee

Let's go back in time again, as we did in Chapter 3, to that period when the human animal was ready for the special gift from the Creative Force—the gift of Mind. It was on the ancient continent of Mu that the human species evolved to its present physical form, and the Manu, or angelic beings responsible for the evolution of man, felt the creature was ripe for Mind, the quality that would set this animal apart from the rest of the animal kingdom.

Since the universe is a huge place, God decided to create by committee, using a standard chain of command. He was boss, then came the archangels and the angels, all taking part in an experiment in "new product development." If you think I'm kidding, you're mistaken; I accept this thesis and intend to defend it rationally. It is certainly as logical as other religious concepts that deal in the metaphysical.

Someone will inevitably ask the question, "Who

created the creative committee?" To that I candidly reply I don't know. But obviously there is something a lot smarter than you and I running this cosmic show, and this being has been doing the job for a long, long time. Science, no matter how sophisticated it has become, still falls short of providing the answers to the "why" of things, and religion, especially the fundamental church-oriented Christianity that so dominates America, is chock-full of irrational thinking. Where does one turn for answers?

I remember asking my father that kind of involved philosophical question when I was 12 years old. We had just purchased a new commercial fishing boat and were bringing it up the California coast from Newport to our home in Avila Beach. The night was clear and stars freckled the night blackness brilliantly. I had been reclined against a coil of line peering up into the sky. The immensity of it all had an overwhelming effect on me. Earlier I had asked my mother about God and creation and religion and such, because we didn't belong to any church and I had no traditional dogma shaping my boyhood thinking. My mother, who had atheistic learnings in those days, looked down at me when I asked if there really was a God and said without a trace of humor: "If there is a God, Tommy, he sure must be a character. Anyone who thinks all the crazy goings-on in this world was 'specially created' has rocks in his head." Obviously my mother was of the opinion that it was folly for anyone to believe that some omnipotent personality had created people and then peered down at them individually to applaud when they were "good" and punish them unmercifully when they were "bad."

My father seemed far more philosophical when I posed the same question to him. Floating in the channel between the Santa Barbara Islands and the California coast late one summer's night, my father—a grammar school dropout, but self-educated man who

82

lives close to nature as a fisherman—answered my query about the mysteries of the universe:

Tommy, the way I see it, the world and the universe and everything are inside a great big brown paper bag. There's an entrance on the right to let living things in, and an exit on the left to let dying things out. Everything comes in on the right and goes out on the left, without exception. But you watch, people will come in on the right and start spinning around trying to find a different way out, or trying to keep from going out at all. Some of them will spin faster than others, and they'll become real experts at spinning around during their lifetimes, but in the end there are no exceptions—in on the right, out on the left. And the only one who really knows what it's all about is the joker holding the bag—and he ain't talking.

In academic terms that viewpoint would be labeled "deism," meaning that the idea of a creator is acceptable, but after the creation he left it all to its own devices, subject to certain unbreakable laws. It took many years for me to fully realize the wisdom of what my father was saying.

Getting back to our critique of how the earth and its living creatures possibly came about, I choose to assume beginnings that make as much good sense as possible. Therefore, I accept cataclysmic geological changes, and a combination of creation and evolution—and we're back to the ancient continent of Mu.

When Mu collapsed it marked the first time a cataclysm occurred during civilized times, thereby becoming the first "recorded" catastrophe. Of course we had no concrete proof of this assumption, but we do have thousands of legends which record the event, and these legends have been a part of human culture since culture first existed. Noah and his ark happen to be one example of the legends.

The submerging of the continent had to play havoc

83

with the usual means of recording events on Mu, so survivors, finding themselves struggling against the elements, could only pass the tale of what had happened to the next generation in story form—and we all know that it takes only a little human embellishment to distort a story.

On the other hand, it is certainly possible that some of the members of Mu's society anticipated the cataclysm and prepared for survival by keeping the knowledge of its history and technology intact. This is not such a far-out assumption, especially when we consider that even today there are organizations that anticipate cataclysmic destruction and are making preparations accordingly. Our own legends, which are compiled in the Bible, tell us of a "righteous remnant" and how it is up to this group to maintain quality when all around are returning to a stone age culture of bash and barter.

According to certain legends, and nearly every legend has some basis in fact, early humanity had it made. You know, the tale of the Garden of Eden and all? I assume such a state existed for early man on that huge continent that is now submerged beneath the Pacific. We are required to delve into some philosophy here because this is part of the "systems approach" of fitting all the pieces of the puzzle together. Our scientists used the systems approach to solve the myriad problems of putting man on the moon. This means they used all the specialists in their respective fields of expertise and blended them together to make sense out of the entire project. In this book we have been blending geology and seismology with biology, anthropology, and philosophy in an attempt to weave a clearer picture of the earth-man tapestry.

I am convinced that each human being is a discreet bundle of mental energy and, as such, was specially created. Why? I don't know, but there is ample evidence of this notion. The first bundles of mental energy

Have you ever thought of what the Earth is like at the center?
Consider the soft-boiled egg concept—a thick, liquid core,
fairly solid middle mass, a mucous layer beneath the shell. This
analogy appears to be scientifically sound and has been pro-
posed by several scientific authorities.

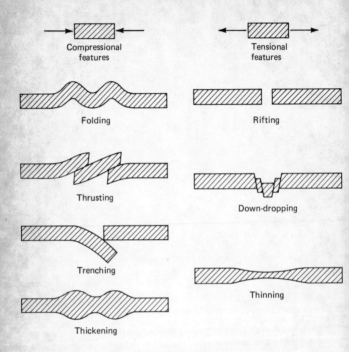

Compressional features

Tensional features

Folding

Rifting

Thrusting

Down-dropping

Trenching

Thinning

Thickening

Here are the various ways seismologists explain the behavior of the Earth's crust as it "drifts" around over the molten magma. These diagrams are adapted from *Scientific American* (November 1971). All the ingredients to make a case for cataclysmology are certainly present, for example—imagine the forces required to cause folding or thrusting!

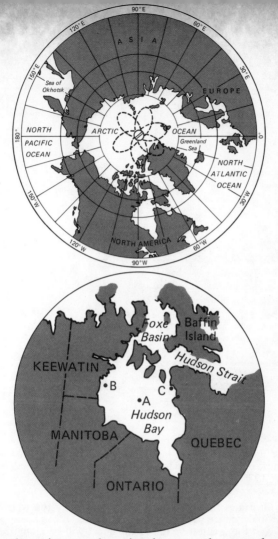

The polar region as we know it today; move the center down to the lower reaches of the Hudson Bay and you can clearly see how an "ice age" occurred in Wisconsin, Ohio, etc. Now, imagine that you are looking down at the pole from outer space and the concept of a "golden screw" marks the spot; as the Earth spins the screwhead would make a cluster-shaped motion due to the wobble as we spin on our axis Below, note pole locations in Hudson Bay area about 15,000 years ago . . . an assumption based on this theory.

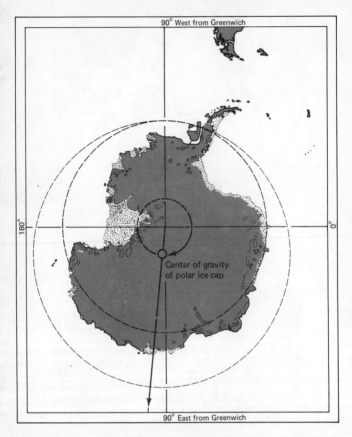

This illustrates the centrifugal effect caused by the weight of the off-balance ice pack covering Antarctica. It has been calculated that the greatest "pull" of the effect will be along the 96 degree East meridian. The pole is located more than 330 miles from the center of gravity.

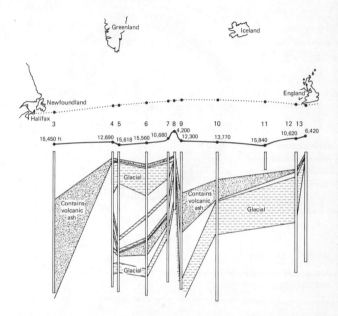

In 1936, Dr. C. S. Piggot sampled the bottom of the Atlantic. The chart above shows the thirteen core-holes. The "mid-Atlantic ridge" shown in the center is called the "Faraday Hills" and significantly the core samples proved that the ridge was once high and dry, separating the sea water. Microscopic flora was found west of the ridge, but east of it only coarse sand and mud identical with glacial moraine deposits. The differences could only have been accomplished if the ridge once reached well above the water level. This information is taken from a booklet, *Atlantis, a Verified Myth*, by Dr. Rene Malaise of Sweden.

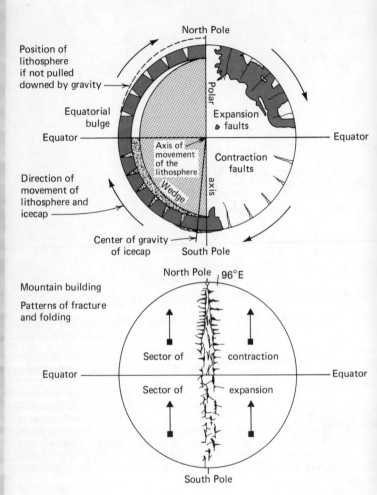

North Pole

Position of
lithosphere
if not pulled
downed by gravity

Equatorial
bulge

Equator

Polar axis

Expansion
faults

Axis of
movement
of the
lithosphere

Wedge

Contraction
faults

Polar axis

Direction of
movement of
lithosphere and
icecap

Center of gravity
of icecap

South Pole

Mountain building

Patterns of fracture
and folding

North Pole   96°E

Sector of   contraction

Equator

Sector of   expansion

South Pole

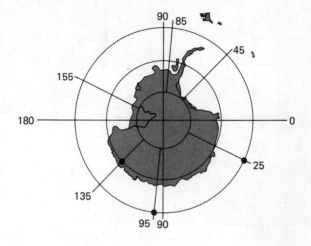

These three illustrations describe James H. Campbell's computations for his theory on the contracting and expanding of the Earth's crust, the Earth's fault system, and Antarctica with positions of the former South poles.

The location of the continent of Mu as described by
James Churchward. Based
upon ruins found on the
Pacific Islands today his
theory may be correct. The
lesson we might learn re-
garding the Churchward
theories is that we are
wrong to toss out an entire
idea because of a few
errors in the presentation.

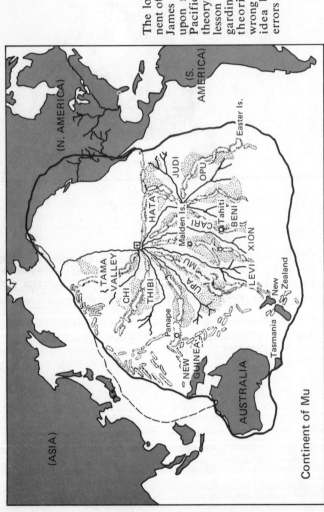

Continent of Mu

Escape from Atlantis. The beginning of a continuous bas-relief frieze discovered by Maler in Yucatan, which suggests to a remarkable degree the Atlantean cataclysm. The above photograph shows a pyramid and temple collapsing, a volcano in eruption, and the land sinking. The figure in the water suggests destruction of life by drowning. Many escaped as symbolized by the figure in the boat.

THE LEMURIAN LAWS

1. No man shall profit at the expense of another.

2. No man singly, nor the commonwealth collectively, may take anything away from another by force.

3. All natural resources shall remain the property of the state or commonwealth and may not be claimed as a personal possession by any individual or any group of individuals not constituting the entire citizenry.

4. Every citizen and every child thereof shall be entitled to and receive equal education, equal opportunity for the expression of his ability and equal standing before the laws of the land.

5. All advancement in position shall be based upon merit and the performance of service alone.

6. No individual shall be entitled to retain as a personal possession anything for which he has not personally compensated in equal value.

7. No individual shall have the right to operate in the environment or personal affairs of another unless asked to do so by that person. The commonwealth or government may do so only where criminal or treasonable intent can be proved, or the civil rights of another have been violated.

8. No one may intentionally kill or injure another person, except in the defense of life or state.

9. The sanctity of the home shall be kept inviolate, and no woman may be taken in marriage without her consent.

10. In all matters affecting the common good, and when no violation of Natural Law is implied or involved, the opinion of the majority shall rule, subject only to the consent of the Elders whose decision shall be final.

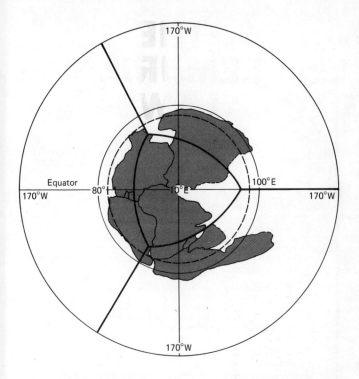

Before the drifting began, the continent Pangaea as envisioned by Dr. Athelstane Spilhaus in the *Smithsonian* for August, 1976. Reversing the present drift of the continents takes us into the past when there was but a single landmass, called Pangaea. The circle in dashes represents 80 degrees from its center, occupied by the land. The solid circle is a part of the grid. The projection of a tetrahedron, in black, is significant because its three points fall on the splits that tore Pangaea apart.

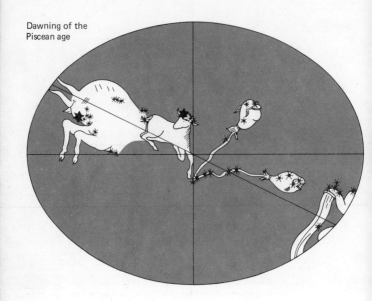

Dawning of the
Piscean age

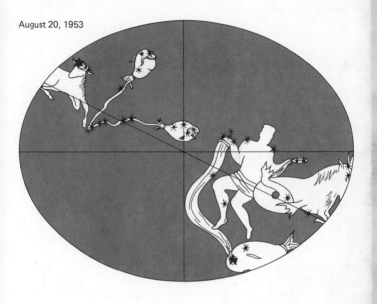

Is there a cosmic connection between astronomy and astrology? Is there a connection between the design of The Great Pyramid and the Birth of Christ? What of the Great Ages of the Zodiac and their influence on Man? Some startling information and ideas are to be derived from clues in these representations of stellar alignments. Can you find the significant pattern in either of these two illustrations? See Chapter 6, The Brotherhood of Man.

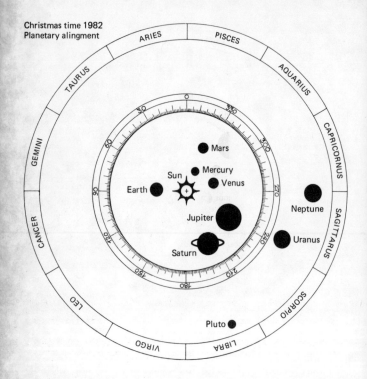

Christmas time 1982
Planetary alingment

ARIES — PISCES — AQUARIUS — CAPRICORNUS — SAGITTARUS — SCORPIO — LIBRA — VIRGO — LEO — CANCER — GEMINI — TAURUS

Mars
Mercury
Sun
Venus
Earth
Jupiter
Saturn
Neptune
Uranus
Pluto

The planetary alignment in late 1982, at about Christmas, indicates a pattern that could cause great sunspot activity and enormous earthquakes.

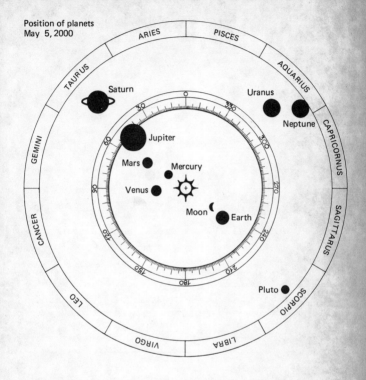

Position of planets
May 5, 2000

The alignment of the planets for May 5, 2000. This dynamic arrangement gives me reason to believe that another great cataclysm will affect the crust of the Earth on that date . . . only twenty years from now (see Chapter 7).

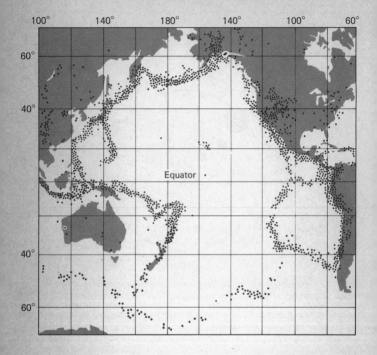

The so-called "ring of fire" in the Pacific, is said to outline the Pacific plate. Since the crust of the Earth is so unstable it is not unreasonable to see differences between today's Pacific plate and the version originally put forth by Dr. Stelle.

(minds) that *incarnated* into the bodies of *homo sapiens* found themselves siblings of animal parents living in an edenic environment on Mu. All the qualities of Mind that make up these "egos" were functioning within animal bodies which had been evolved by the "forces of nature," meaning forces inherent in the environment purposely designed by a higher intelligence.

Food was exceptionally abundant because weather conditions were more perfect than anything we can find on earth today. We assume this is true because of the persistence of the legend to that effect, and because a huge continent such as Mu could readily attain optimum climatic conditions year-round.

It wouldn't require much readjustment of the forces to have such optimum weather conditions. Something the size of Mu would surely have had a large area where rainfall and temperature were ideal. Southern California comes close to that ideal today; a lack of rain and too much fog are the only drawbacks. Climatic conditions in America have a great deal to do with its rise to world leadership in the international economy. Look at the globe of the earth and you'll see that the United States is the largest body of temperate land ideally situated for weather currents from one ocean to another. No other farming area in the world can grow the staple crops on rainfall alone that are produced in the American Midwest. The idea of a Garden of Eden is certainly not far-fetched.

There are many versions of what happened to those early egos who inhabited the Garden of Eden, so I might as well toss in my own which is taken from *The Ultimate Frontier*. According to this book, the purpose of human life is to build character and grow in knowledge until becoming the relative equal of the Creator. All the tools are readily available to every ego for this building program because every discreet mind has the qualities of memory, desire, will, curiosity, consciousness, conscience, creativeness, intuition, emotion,

85

and reason. *The Ultimate Frontier* stresses the reality of "planes of existence" and the "reincarnation" of minds into human bodies over and over until the character, or soul growth potential, is attained and the mind is "liberated" (this is a Buddist term meaning no longer needing to reincarnate). Those ten qualities of mind are identical to the qualities inherent in the Creator of Creative force, thus "we are created in His image."

Evidently we were supposed to start learning and growing from "clods to gods" simply by inhabiting animal bodies and going about day-to-day living. Essentially, the only difference between human egos (you and me) and the Creator is the ability to always make the "right" choices, always think the right thoughts, and always bring about the right consequences (although) I can't even honestly define what is meant by "always right."

One can imagine what it was like for the first *minds* to be born to animal parents. Talk about a generation gap! Mama and papa are brutes with only animal cunning and instinct built into their bodies. Junior is also brutish, but has the ability to reason, to be curious, to be intuitive, and so forth. He would learn very early, for example, that bananas were easier to chew when peeled. What a life! Grapes and bananas and papaya and sunshine and clean air and water and girls and dumb parents. No wonder it is said the angels were disenchanted with the outcome of their part in the creative process and started bickering about humankind and the lack of character growth.

According to one version of this story, the Angel Lucifer and the Angel Jehovah disagreed as to how mankind should develop character; when Jehovah wasn't looking, Lucifer made his point by taking away the edenic conditions. Now, that legend could very well be a colorful way of saying that the Higher Intelligence that is running the inside of our brown paper bag wanted mankind to struggle a little and be forced to use his mental qualities to grow instead of lollygagging in Eden. Fundamental Christians believe the Adam and Eve version which transfers the blame of man's moral frailties.

All those stories are probably a way of describing a "minor" cataclysm that came along and interrupted the climatic conditions responsible for the plush situation at Eden. With the edenic conditions destroyed, mankind was forced to pit his wits against the elements and learn to use all those mental tools the Creator gave him. Some people, especially those in Lemuria, praised Lucifer for taking away the easy life and helping them work from clods to gods; however, more people through the ages have damned him for it. We now blame him for all our problems and he's known as "the devil," "satan," or whatever.

This concept of "fallen angels" had to come from somewhere, and I'm sure most scientific atheists feel as I felt when I was among their ranks (that is, the ranks of atheistic anthropologists, not the ranks of fallen angels). The scientific type tries to conceive how early thinkers came up with such a "strange" idea as fallen angels or Manu. The usual solution is to figure that people generally want to bring everything down to a more casual level, a level easier for coping, so somewhere along the line the perfection embodied by the concept of angels had to be dragged down. Having a cynical view of human behavior, it would be an easy matter to accept that notion. However, it is inconsistent. *Every* legend—at least every serious legend—has some basis in fact so far as I'm concerned. Those facts might not have occurred on the physical plane and this can pose a problem, but the facts did occur and reports of them were somehow handed down until they were made a part of physical man's information bank.

Dr. Stelle's book tells us that early civilizers had help from two angelic beings from other planets in our solar system because the angels assigned to earth were all shipped out by higher powers, due to hassles over the edenic state. This version is also presented in *The Ultimate Frontier*. At any rate, the two higher beings, the Manu for those brutish-looking humans who were seeking a better way, were said to have come from Venus and Mercury. In essence what we have in this story, this religious concept, is von Daniken's idea, only on a much higher level. Some things begin to make a little more sense with this viewpoint.

Having been told that Manu from Venus and Mercury were part of the formation of the first civilization, later humans found ways to dramatize the story. In the glory days of Lemurian civilization, it was probably pretty accurate, but the hand-me-down versions that followed the cataclysm became distorted and we find yarns like those about Zeus tossing thunderbolts, engaging in love affairs with mortal females, and so forth. There is a so-called Venusian calendar of 225 days found carved on stones associated with Lemurian colonies in South America. In light of Stelle's story, that carving makes a little more sense. Venus was special to Lemuria and was watched closely. What other reason would people have for carving a "Venusian" calendar in earthbound terms of twenty-four-hour days?

Imagine how this is going to irritate von Daniken—someone actually saying that his ancient astronauts were not physical, but ethereal; angels yet! The von Daniken followers would rather have physical spacemen come to earth thousands of years ago to cohabit with the brutish populace and bring about the genius we find carved on the rocks of antiquity.

When it comes to genius, humans have all the equipment necessary in those ten qualities of mind, and outer space people have enough problems of their own without coming here to fret over ours.

It was a cataclysm that snatched the Garden of Eden away from us and replaced it with winters in Chicago.

It may have been accomplished with the grim approval of Lucifer over the objections of Jehovah, we really cannot say for sure, but it makes a nicer story if we go along with the idea of an army of angelic beings helping out the Creator. Actually, it's a nice concept. Do you have anything better to do with your lifetimes than work to become the equal of your Creator? Well, that's what this colorful mythology is all about; and it might not be far-fetched mythology at all. I've always been extremely curious as to the "why" of it all. Did your children ever ask, "Why am I, Daddy?"

We have a pretty good idea of what we are and how we work; we can pretty well tell where we are and when we are—but I know of no one who really knows why!

Back to evolution and creation. The huge continent offered more than enough room for developing skills and overcoming the elements. For several hundred thousand years, man slowly developed, pretty much as described by anthropologists I would imagine. We were fearsome and fearful creatures, and the going was slow, often violent, generally ignorant. A few individuals came along periodically who used their minds rather well; this is a thing called genius. And they would detach from the tribal pack to set out in search of the better way that intuition told them existed. These exceptional thinkers managed to find one another and learned the value of cooperation at an early stage. Dr. Stelle points out that the founders of our first civilization were aided considerably by these unique individuals. He claims they were called "elders" by the forest, cave, and plains dwellers who populated most of Mu in the stone age tribal cultures.

Finally, according to Stelle, two young men of the Mu-yans tribal group became inspired by the elders. They had the leadership abilities to bring together the best of the three types—forest, cave, and plains peoples—to forge the beginning of civilization: the be-

ginnings of praise and joy at accomplishment, the beginnings of cooperation and industry for the common
good above and beyond primitive tribal loyalties. We
certainly didn't need any outer space connection to find
those qualities; we merely had to bring them to the surface from within. We had to overcome fears and suspi-
cions.

I use the pronoun "we" when talking about the humans who existed on this ancient continent. I accept
that it was indeed "we," by virtue of acceptance of the
theory of reincarnation. We have indeed all incarnated
thousands of times since the joker holding the bag decided the bodies were ready for mentality more than a
million years ago. Though you may not personally believe in the theory of reincarnation, don't let that deter
you from reading on. This book is not trying to prove
reincarnation, but it does suggest that there is substance behind certain myths.

Legends and myths are in themselves fascinating
studies, and they can provide us with a tremendous
body of evidential material for conjecturing about what
mankind's prehistory was really like. Dr. Stelle's map
shows a dozen very large geographical-tribal areas
making up the larger portion of ancient Mu. The first
civilization began on the Rhu-Hut Plains, observed on
the northeastern part of the continent. According to his
esoteric information, the huge valleys were named as
follows: Tama, Chi, Thiba, Upa, Mu, Levi, Xion, Cari,
Beni, Hata, Opu, and Judi. Now this kind of information is the purest of conjecture, but since Dr. Stelle's
map proves to be so accurate tectonically, why not
consider the rest of his information valid until proven
otherwise? An interesting tale comes out of the
teachings of Dr. Stelle which has ramifications even today.

Two of the twelve tribal groups of man (does that
have a familiar ring to it?) refused to cooperate with
the other ten tribes as the first civilization was getting

on its feet. Hostilities broke out and finally the two tribes that chose not to go along with the others were defeated. As punishment, they were banished from Mu and forced to wander over other parts of the world. The Lemurian philosophy handed down by the followers of Stelle teaches today that the Archangel Melchizedek is the "Christ" entity most involved with the character development of mankind on earth. Melchizedek has "come down to earth" to work with humankind on three occasions so far, and he is being counted on for one more appearance. He ruled ancient Lemuria during its height and is said to have been fully responsible for the decision to banish the two tribes of Levi and Judi. Just those two tribes were separated from the original twelve, so it is not the "ten" lost tribes, but only two. Hence, thousands of years later, during his "third" appearance among mankind as Jesus Christ, Melchizedek *chose* to work his plan through the remnants of those two tribal groups. It is said by many Stelle followers that Melchizedek also chose to sustain the physical pain and torment of the crucifixion to pay "karma" for his decision for banishment.

Not only is this an interesting and theologically plausible new twist on the idea of the "chosen people," but it also offers an interesting angle on the "avatar" or "returning diety" concept that has held sway over so many for so long in the Orient.

Another curious bit of legendary conjecture stems from the tale that the men who became the master stoneworkers of Mu adopted special rituals and traditions and called themselves the most loyal sons of Mu. Stelle pointed out that the ancient Lemurian language was quite similar to much of today's English, which has slowly been reevolved by more advanced individuals taking part in a deliberate plan. Thus, this group of skilled artisans came to be known as *Musons*. Also interesting is the fact that the Masons themselves cannot be certain of their own origins.

91

Those twelve geographical divisions on Mu were actually huge tracts of land. Based on Stelle's map, we can see, for example, that all of Mexico and Central America would fit easily into any one of the huge valleys. Such geographical division, we may assume, could well be responsible for the so-called "racial" differences in man. The origins of race are unknown and puzzling to anthropologists. There are a number of elaborate theories about how races came about, but nothing any more concrete than the thinking that I'm about to share with you. The question is, how did so many subspecies of *homo sapiens* come about?

First of all, it is important to realize that race is purely physical. Our social conditioning might not allow all of us to view race as such. A human mind may incarnate into any species and the identical qualities of mind will be present, so in essence there is no difference in humans. Yet, there is a broad spectrum of physical differences that appear to be greater than they really are. The physical differences between humans are minimal; the physical similarities between humans are maximal. Though every individual is readily distinguished from others, there is no difficulty whatsoever in always identifying the species. I like to think of our similarities this way: If people of different races were truly different, like dogs and cats, for example, then they would not be able to reproduce—like dogs and cats. There's no problem whatsoever in two humans of different races getting together and making babies, unless it's social conditioning.

Racial characteristics do not require eons of time to develop as you might think they would. For example, in less than two-hundred years the American black has evolved an almost "totally different race" than the nilotic negroid and forest negroid stock in Africa. The major physical differences of skin pigmentation, hair variety, and skeletal stature probably evolved among the people during the million or more years in which they had "egos" incarnated into them. My reasoning for this assumption is that only a "thinking" creature would consciously choose to segregate because of apparent physical differences. The black leopard, for example, doesn't have a mind, so it isn't conscious of its physical differences from spotted leopards. Therefore it hasn't pulled away from, or been ostracized by, the other leopards. If such a division occurred, you can be certain that it wouldn't take long before black leopards were a dominant strain with occasional spotted offspring.

Look at it this way: Whenever a human animal was born with additional pigment or kinky hair, or was an albino, that individual would be conscious of that difference and would tend to seek out others more like himself, thereby bringing out racial changes through selective breeding. After all, would you want your sister to marry one?

Physical differences between people are accomplished by less than 5 percent of the genes which determine body characteristics. It can be clearly shown that nutritional factors alone can cause physical changes and eventually racial changes. To clearly illustrate how minimal racial differences truly are, I enjoy citing this example: The darkest skin pigmentation—the blackest skin on earth—belongs to the "white" race. That's right, the *blue-black* Melville Islanders, an Australian aboriginal group, have the darkest skin, but are caucasoid in that they also have blue eyes and red hair. That tidbit of anthropology should be given to all the bigots to chew on.

An interesting aside to this digression is that part of Dr. Stelle's esoteric information relates how some of the ancient Lemurians took exceptional pride in their racial characteristics, and how at one time the government encouraged the development of a special breed of distance runners who were practically "blue" skinned. Could it be that some of today's Australoids are an offshoot of that ancient strain? Aboriginal runners are famed for their ability to run or lope along for hundreds of miles at an incredible pace.

All this talk of race brings up another question that has perplexed science for years and years: What about the missing link? In anthropology the "missing link" is said to be that stage of evolution which links man and the apes. The term "missing link" is a remnant of the older paradigm even though it still finds use among laymen. Today it is assumed that man and the apes both evolved from the same source—probably a tree shrew or something of that ilk. That's merely an assumption though, and even the most respected anthropologist in the world won't try to tell you that it is anything else.

Archaeologists have found bits of human skeletons dating back nearly three million years, so the length of time our species has been around has been continually expanded in the last twenty years. However, there are two things wrong with such assumptions. First, since earthquakes and volcanoes and cataclysms are a reality, how can the archaeologist be sure the skull he finds 300 feet deep in strata wasn't thrust there at a later time by some kind of upheaval? I understand there have been considerations made for this, so it's not nearly so big an objection as the next one. How is it possible to tell if the ego had incarnated into the soul by examining bits of a million-year-old skeleton?

Am I being facetious? Not in the least. I'm trying to make a point. That point is: We cannot study or theorize about our dim past or the beginnings of life on earth without full consideration of metaphysics. Things like the akashic record, human intuition, and so forth cannot be ignored—they are a part of this story. In fact the metaphysical is probably more important than the physical in the final analysis.

Up to this point we have been busy with musings about what may have truly been "the how" in how civilization began. We have considered all kinds of evidence for past "great" civilizations, but we have not considered what it is that a civilization is supposed to do. Certainly, we are all capable of asking, "What's in it for me?"

And we get right back to that question the kids ask before they finally learn we don't have the answers and shut up. Why? Even if the edenic state was lost and early man had to struggle to survive, why should he and his neighbors bother to civilize? What is needed besides cooperation if survival is the goal of the species as biologists contend? It is certainly possible to cooperate without civilization; we can be sure that cave men cooperated to rid their houses of whatever brought them danger. What is it that compels humankind to seek more than other biological organisms seek out of life?

What's the matter with a cave and a fire? Why seek walls and marble columns and rugs and linens and tapestry and decorated pottery and tailored clothing? None of these things are demanded for survival. The human species is one of the largest and most ferocious animals in the animal kingdom—we could have survived, indeed we could have dominated without all the frills. Put a big rock or branch in the hands of a 200-pound football player and he's a match for nearly anything in the jungle. Besides, animals don't go around killing one another for the fun of it—they don't have consciousness; it simply never dawns on them.

Consciousness is the key word in the story of man on earth. No matter how complex the helixes get in the spirals of DNA, there is no way that atoms can so arrange themselves that they are able to perceive of themselves and consider what action they will take by means of chemistry alone. The sciences of microbiology and molecular biology can show us how utterly fantastic and wonderful the natural chemical system is, but once again we see that the physical is only a part of the answer. The role of the metaphysical is obvious, but mysterious and sometimes scary.

Consciousness has been debated by dusty old philosophers since minds incarnated into the species. The best way I have found to define consciousness satisfactorily is to use my father's method of comparing humans to animals. Does your poodle know there are stars in the cosmos; does he care? Does your poodle know there is more water than land on this planet; does he care? Does a sheep give a hang about why he's being sheared? Does a fish comprehend that he's at the end of his rope when he feels the hook? Though you may think your pet is "practically human," the answer to all of the above is "no." Humans are the only animals who "think" and, of course, cynics will chuckle at that one.

It is apparent that this thinking process did not evolve, though science is still searching for evidence to support a theory of evolution. From whence did it come? Now you can see that such a question brings us right back to a million or so years ago when those discreet bundles of mental energy, specially created on the fourth plane of existence, were given the green light to begin incarnating. That's the way I *believe* it happened, but proving it may be nearly as difficult as what scientists are trying to prove, though not quite.

Those ten qualities of mind can be a curse as well as a blessing, obviously. The joker holding the bag didn't spell out in black and white exactly what every ego was required to do with the thinking process in order to find perfection in his daily life. That would have been the easy, comfortable, and worthless method and would, quite probably, take all the fun out of creating. After all, the Creator could potentially have 13 billion "puppets" on a planet and just pulling the strings could wear even God out! (According to esoteric information there are a total of 13 billion egos created for the planet earth; however, under normal conditions, only

96

about one-sixth are incarnate at any one period of time. The coming of a cataclysm usually signals a need for more souls to insure survivors; hence an increase, or what we term "overpopulation.")

The purpose of civilization is to construct an environment that is conducive to character or soul growth —and it's really that simple. Our species has a built-in mechanism that seeks comforts, refinements, and leisure in order to accommodate our need to ponder this purpose and exercise consciousness to its fullest potential. Dr. Stelle taught that the first civilization on Mu—which lacked a history of sophisticated bad example from previous civilizations which might have deterred it from its path—rose to fantastic heights in promoting character growth for its citizens. According to his philosophy and that of many organizations around the world, each of us has the potential to advance by leaps and bounds if we truly desire to do so. As one practices the *virtues* they soon become habit, and the ego soon learns he has control over himself and his environment. In the case of humankind, the powers inherent in mind come to the fore and each individual begins to view the other levels of existence more clearly (called clairvoyance) and understand the true purpose intended in the great plan more thoroughly. True civilization helps bring such individual growth about for all its people.

What does our twentieth-century civilization promote? We are part of a society that has evolved freedom of movement and thought, so we aren't doing too badly in light of the past 7,000 years of recent history. We have a myriad faults and problems, but nothing that couldn't be straightened out with *genuine* effort on the part of *everyone*. If what I'm suggesting is true, then it's time to get back on track and begin to get out of life what was intended!

According to the story, the civilization on Mu did a good job and lasted for 52,000 years. There was a period during this time when the archangels incarnated, served the people directly, and the soul growth rate was terrific. Although such a concept may strike you as ludicrous at first thought, it's not as silly sounding as the predominant belief in Christian America today. Most believe that God had a son who came down from the heavens, tried in a lifetime to show by example the proper way to live, and was rewarded for his effort by being crucified. Just think a moment—isn't it every bit as feasible to have a High Being incarnate into physical reality and rule over a citizenry worthy of such rule as it is to have a High Being come into reality and be crucified by a bunch of neurotics who were afraid of being found out?

One predominant Christian theme today teaches that the Higher Being (Christ) is indeed going to return to physical reality in the near future to preside over "heaven on earth." There is some controversy over precisely what makes up a "heaven on earth," but this too is essentially the theme of this book. We should remember that today's Christianity is based on the same set of beliefs which govern the present historical paradigm. Since it is generally believed that man has dwelt in civilization for only around 7,000 years, it is generally *not* accepted that the Christ entity had any need for earlier appearances.

Getting back to the march of time and mankind, *The Ultimate Frontier* describes the task of civilization as follows: "The only real purpose of a civilization should be to promote the development of advanced egos. Peace, prosperity, freedom, and happiness are important because these conditions enhance spiritual advancement. But without the proper philosophy to guide men there is no progress. Truth is the *seed* of advancement; a sound, serene civilization is the *climate*; and man's mind is the *fertile soil*—together they give rise to human perfection."

"... it was in the formative years of the Lemurian

Empire that the seeds for its destruction were sown."
In a nutshell the "seeds of destruction" consisted of allowing great numbers of new people to come into their
civilization to enjoy its advantages without taking care
that they also accepted the attendant responsibilities.
Soon, the story continues, the original citizens became
alarmed at the number of laborers who enjoyed the
fruits of civilization but didn't take the time to look
into the true purpose of its inception—soul growth.
The government attempted to implement an education
program, but found to its dismay that the educators
soon found themselves in a powerful, self-satisfying
position—they were looked up to by adoring crowds
as they preached the philosophy and a priestly class
was born.

As a myriad examples in today's society show, it is
very easy to be swept away on a wave of self-impor-
tance when people flock into your church once a week
to hear you tell them of the advantages of good, the
detriments of evil, and the real purpose of all creation.
You become a special envoy who has the answers, and
that's bound to bring on a nice feeling. But the priestly
personalities who came into being on Mu, people just
like you and I, soon learned that if the *truth* of the philosophy sank in, they would soon lose their flock.
You see, the *truth* is that people only learn by getting
out in the world, accepting responsibility, and developing by practice and hard work on a daily basis—
grow by doing, not by occasionally listening to someone else talk about it.

Two wrongs can, in fact, make a right if results are
the only things that count. The Lemurian priestly class
which developed for the wrong reasons worked to retain their audiences, and the audiences wrongfully
lapped up what the priests had to offer. Today we are
still dominated by that syndrome and, for all practical
purposes, the majority will proclaim that it is right.

The comforts of the Lemurian civilization were

99

many, and soon the original idea of using leisure time to develop one's own character became boring. This is an attitude that is rather prevalent today because, as is obvious, we humans are entitled through the gift of free will to use the qualities of mind in any manner we see fit. The foibles and follies of mankind are due to the combination of free will, ignorance, and sheer laziness. It is easier not to exert oneself.

According to *The Ultimate Frontier,* problems grew in spite of the advantages of civilization:

Before long there were more noncitizens than citizens in the empire; but because the citizens were those striving on the path to immortal perfection, the power inherent in their aggregate advancement far outweighed the lesser masses. . . .

The laborer class could not comprehend the religious philosophy of the citizenry and therefore was not attracted to it. In hopes that the laborers could be enticed to study Universal Law and come to an understanding of the vast advantages of striving for perfection, the citizens set up "churches" for the laborers. Breathtakingly beautiful buildings were provided, rituals employing fascinating symbolism were instituted, and the church leaders were provided with magnificent robes and other trappings. The laborers eventually were attracted in droves for much the same reason that people go to a parade or circus. The plan was to arouse the curiosity of the laborers and then unobtrusively implant the desire and incentive toward concerted soul uplift.

The plan was a miserable failure; for not only didn't the laborers seek to understand and advance, but with the passing of centuries, the church leaders succumbed to the delight of being literally worshipped by the laborers. The nation was so abundantly prosperous that profit was not the motive of those who later became known as priests; it was public adulation and the delicious control over others that made these priests seek ever greater power for themselves. Noncitizen laborers were attracted by the priests' promises that they would return the empire to edenic paradise if the priests could achieve control over the government. Not all the laborers, however, were taken in by the priests' claims, but these exceptions were in the minority. The priests promised that no one need ever work again if Eden could be reestablished. All that men would have to do is to pick food from the nearest tree and have all day free to play or rest. The unthinking, impractical believers of this vision of paradise were told by the priests not to try to reason out how this return to Eden would be accomplished, but merely to put themselves into the hands of the church and have faith that it would work. The highest motives of spiritual good were mouthed by the priests, and for hundreds of generations their followers clung fanatically to this dream of endless bliss. Every effort by the government to stamp out the perverted lies by which the priests enthralled their followers only made martyrs of the priests and drove them underground. The priests promoted sedition, and their followers were eager to die in defense of their priests and lofty ideals. These impractical laborers gave themselves the name *Katholi* and believed themselves to be highly spiritual and idealistic.

On the other hand, the practical-minded noncitizen laborers realized that a return to edenic conditions would put an end to civilization and all its advantages. The Lemurian Empire enjoyed an abundance of labor-saving appliances and luxuries which are beyond your present understanding; yet, the Katholis had been duped into wanting to trade civilization for a labor-free paradise. The priests had freely implied that none of the material comforts would be given up in paradise and that their god would provide everything for the Katholi believer in return for his true worship. Because the great Angel Lucifer, had been responsible for the abolishment of Eden in order that men could begin on the road to spiritual advancement, the Katholi were led to believe that this

101

angel was the most loathsome ego ever associated with the earth.

Those laborers who couldn't swallow the idea of a paradise replete with endless luxuries and civilized conveniences without anyone working to manufacture these commodities or to provide public services gravitated to an alliance with the citizenry of Lemuria. The noncitizen laborers who could not tolerate the priests of the Katholi churches banded together and called themselves the *Pfree*. They were usually craftsmen and highly skilled metalworkers who enjoyed their skill in building things. These practical laborers even adopted Lucifer as their patron in order to strongly differentiate themselves from the Katholis. Their natural ability to become foremen and leaders in their crafts brought them into close contact with engineers and administrators who had the education entitling them to citizenship. This contact provided a natural opportunity for the Pfrees to learn the advantages and philosophy of the citizenry.

The idea that men might enjoy a paradise of plenty without effort is absurd. The angels provided every possible raw material which man can use to further his understanding of the physical plane, but not until man expends effort upon raw material can it serve him. In this world there is no such thing as something for nothing! Man has a built-in desire to visualize and to achieve goals. To achieve anything necessitates the expenditure of human thought and energy. When a man is working toward a goal, he is happy; when he has no goal, he becomes dissipated and feels cast adrift. To build and to create are the bases of man's joy. To build is to bring wealth into existence.

The Katholis and the Pfrees came to be poles apart in their philosophies, and yet, both were wrong. Each faction of laborers lacked the citizen's proper balance between blind faith and skepticism; the Katholis prized spirituality and the Pfrees practicality. The obstinate onesidedness of each group of laborers inevitably led to open conflict among them, much to the distress of the citizens and to the detriment of the empire. Finally, the citizens began a movement to encourage the emigration of the Katholis to hitherto unpopulated continents by offering extraordinary inducements, and hundreds of millions of laborers became enthusiastic enough to colonize other lands. The thought of being able to found their own nations according to their religious beliefs strongly appealed to the Katholis. The principal Katholi colony was India where the Katholis' priests established a hierarchiacal rule over the settlers and readily enslaved them. Meanwhile, multitudes of Pfree laborers set up their own nation on the Poseid group of large islands in the Atlantic Ocean. . . .

India still bears the scars of those thousands of years under the domination of priestly types; only in India has something as tragically incorrect as the caste system been able to evolve and sustain itself.

The wars between Atlantis and Rama have come down to us in the form of myths about the gods bickering among themselves. Atlantis and Rama were the dominant civilizations for 18,000 years until Atlantis submerged in an isostatic crustal adjustment. Ancient Rama was not wiped off the map by cataclysmic forces; in fact India is a land filled with ancient and mysterious artifacts testifying to the validity of this thesis. The Raman civilization destroyed itself slowly by eroding away the human qualities of its citizenry. It is significant that many ancient civilizations, including the Egyptian and the Ulmec-Toltec-Aztec of Mexico, deteriorated rather than improved with age. I am convinced that this curious deterioration of a culture is caused by people giving up their full responsibility to a few individuals in return for "leadership." There is an ominous presence of just such a social phenomenon alive and growing around us today.

Thus far we have covered the "prehistoric" times and perhaps have conjured up interesting notions about how so many "unexplained" legends might be explained. But, a key question remains: If the Pfrees and their practicality went to Atlantis and the Katholis took their spirituality to India, where did the nicely balanced, advanced citizens go when Mu vanished?

CHAPTER 6

The Brotherhood of Man:
Playing God and Working at Mankind

We've all heard the expression "the Brotherhood of Man." The term can have many meanings, so I'll give you mine. It obviously refers to the *fact* of human equality insofar as our mental qualities are concerned. Also, it most likely refers to the task we humans have before us—the common task of advancing by accumulating knowledge until we are the relative equal of our Creator. That seems to me to be a fair price to pay for the immortal gift of ego, mentality, and personality.

This common bond has led to the formation of a society of more advanced egos, which consists of an equal number of male and female members at all times, I am told, calling themselves the *Brotherhoods.* Those citizens of ancient Lemuria who advanced rapidly toward human perfection by concentrating on developing their character and growing away from the physical toward the metaphysical with *knowledge* soon noticed the plight of the rest who have not been advancing at such a pace. Because an advancing person develops the natural urges of love and compassion while controlling the animalistic urges of anger and fear, it stands to reason that such a person would desire to help his fellow man.

According to my information, and it doesn't vary much from one esoteric group to another, advancement comes in recognizable stages; once an ego is firmly on the path, there is no doubt in that individual's mind that he is on the right track. I think this is what is meant by "finding yourself," an expression heard often today but rarely understood. Actually it is more like "making yourself," because advancement isn't something you bump into by accident; it is something you must develop *entirely on your own*. To me, it is ignorance of this simple truth—that advancement is a reality and must be worked for—which causes humankind the most pain as time passes by so relentlessly.

According to my source, *The Ultimate Frontier,* the stages of advancement range from an Initiate, first degree, through a Master, twelfth degree. An Initiate is an individual who has a firm control over himself and his environment by virtue of his ability to practice the Twelve Great Virtues, regardless of the activities and attitudes of others around him. According to the legends, the citizens of Lemuria during its heyday all practiced these virtues very well. The virtues are simple, but not so easy to "ingrain into habit" as we must do to advance. They include, patience, tolerance, forbearance, devotion, courage, sincerity, kindliness, discrimination, charity, efficiency, precision, and humility. You can imagine what a civilization the citizens of Lemuria evolved!

When a person habitually practices those virtues and goes along in life unobtrusively uplifting everyone around him, he finds that he becomes aware of more than just the physical nature of his surroundings. When he converses he understands more than the mere words; his intuition informs him accurately of deeper meanings. When he perceives, he "sees" more than the physical reality; he sees the vital and astral levels simultaneously. And he can function on the astral level while remaining fully conscious within his physical vehicle. He is a "controlled clairvoyant." Naturally, such a person has a balance between his positive and negative *karma*, meaning he has balanced out his bad deeds, performed during less knowledgeable times, with good deeds during more advanced times. This person is ready for *first degree*. Can you imagine what twelfth degree must be like? It is said that a Master is totally in control of the first four planes of existence and his vehicle is on the mental or fourth plane; he may choose to, but need not incarnate on the physical level. Masters have it made! According to *The Ultimate Frontier,* "all the human beings occupying this planet came into existence at the same time—somewhat more than a million years ago. Collectively, all of these persons are known as the 'human life-wave.' " When the time for change of life waves comes along, everyone who has attained Mastership will automatically move up a notch to the Fifth plane, or Angelic level in the terminology of my faith. Those who have not attained Mastership remain discreet bundles of mental energy, but all memory of this life wave is wiped out and they will begin again, in ignorance, to reach Mastership in another life-wave.

From this point of view the atheist is partially correct. If memory is wiped away when it's all said and done, then nothing has happened. It's all in how you look at it.

The purpose of this explanation of my religious convictions (and obviously these are not scientific facts that I can prove) is to give perspective to the thesis that such advanced individuals not only exist, but they are working hard to help us learn to help ourselves and have been doing so since they saw the handwriting on the wall in the days of Lemuria. Members of the Brotherhoods, people between First and Twelfth Degree, working under the direction of the Archangel Melchizedek (whom we know as Christ) laid out a plan to help mankind evolve another civilization like that enjoyed by Lemurians at their pinnacle. The refinements and beauty of such a civilization, I am told, are beyond the imagination of most of us today. Certainly such a thing may be construed as "heaven on earth."

If all this is true, you might be tempted to ask why such a civilization is so hard to envision; so hard to understand or believe?

I can't claim to be able to answer that question—people evidently have a penchant for making advancement difficult for themselves. Good old ignorance and free will are again the culprits, I suppose. And, there are other factors that seem to shed some light on why it has been so difficult; on why we have had Hitlers in power and insanity as part of our recent history.

The universe was created in accordance with a set of rules and regulations called Universal Law and even in our ignorance we can see evidence of its inviolability surrounding us everywhere. Evidently this is a learn-by-doing universe, and by that I mean that it doesn't matter what I'm writing at this point, you and I will learn of Universal Law only firsthand. This is where *karma* comes into play. Karma is a sanscrit word meaning "carry over," and therefore it is an apt word to describe the basic law of cause and effect. Everything we do "carries over" into everything else. For example, if we extend something such as love from ourselves to another human being, an equal amount of that same something comes right back. It's so simple in essence but most of us have found ways to make the concept complicated.

One of the regulations, apparently, is that every ego must advance *entirely on his own.* Now that means what it says. Jesus won't do it for you for the simple reason that you can do it for yourself. We are all created equally and we have got to use our qualities of mind, which are identical to the qualities of the Creator. We can share information and we can wish each other well; we can even lead cheers—but in the final analysis everyone, even mama's favorite son, must advance on his own merit and effort.

This seems to be one tough concept to get across to people so I repeat; it is this singular factor, namely ignorance of the responsibility of every individual for his own advancement, that has most affected man's history. It isn't the wars, or the political boundaries, or the types of culture, or the political promises in man's history that have had effect. Those are truly insignificant. They are symptoms of an underlying cause. The *cause* of more social problems and political hangups in our last 2,000 years of history is *ignorance* of the *do-it-yourself* concept.

Demagogues could not come along and promise the world to people who understood that the purpose of life was to advance as individuals by means of individual hard work and striving. Government can't do it for anyone, and the old saw about the government that governs least, governs best is absolutely correct. People must first govern themselves in their daily lives. The only reason political organization is needed is to enhance cooperation and improve environment so everyone can practice character growth in freedom and a conducive atmosphere.

There are also apparently regulations against "playing God," even when one is advanced so far above the average that he would appear to be God-like. This means that the Brothers cannot reach back and take egos who are less advanced by the hand and lift them up in the rapture or whatever. Every single ego must put it all together for himself, by himself. It really helps to have full cooperation from others and that's what civilization is supposed to be all about. And we come back again to that idea of a "Great Plan" for the ages—the plan of the Brotherhoods to "guide" us back to the truly "good old days" of ancient Mu.

The concept of a Utopian civilization destined for the future is carried forth by the Christian notion of a "kingdom of God," or perhaps more accurately, a "nation" of God. I happen to believe very strongly that such a civilization is within reach, that it is just around the corner, and the plan of the Brotherhoods is working out, despite a concerted effort against it.

Why a concerted effort against such a good idea? The answer to that question is this thing we call "evil." Here's a word that can really do a lot of damage because it can be used to strike fear into the hearts of men and thereby control them. "Fear no evil" is good advice, but it has seldom been heeded. If advancement by individual effort is our purpose, then we have nothing to fear—*ever*. Fear opens us up to influences that take us away from the concept of advancement more than any other emotion, except perhaps that of hatred. It is evil to knowingly attempt to prohibit or inhibit advancement opportunities for others. Evil grew historically because of ignorance and free will.

The concept of the "devil" is one Christian idea that I have always doubted. There is no specific individual named "Satan," no "fallen" angel out to devour every last soul as far as I can ascertain. I'm of the opinion that this concept is twisted, as many of the original concepts have been; however, I do accept the notion of "evil entities" working to impede the advancement of mankind. Not fallen angels, but fallen egos are involved. Just as the Brotherhoods are striving to help us without interfering, the "evilists" are striving to interfere without helping. In an abbreviated form, here's how it works: Every ego has a mind and that means power from the fourth plane of existence. That power is there for everyone to draw from, but how it is used is up to free will. It can be used to deter advancement as well as to learn perfection. We may have been created equal with regard to receiving qualities of mind, but we have not remained equal since the first incarnations (or beginning of this life-wave). Some people have advanced dramatically and *know* what this journey through lifetimes is all about; most of us have not advanced, or have advanced minimally relatively speaking, and don't know. And there are some who have learned to use mental power as if they had advanced, but actually they are traveling in the reverse direction—these are the evilists.

Just as the Brotherhoods have roots in the wonder of ancient Lemuria, the evilists have roots in the conflicts of ancient Lemuria. The "priestly classes" learned well the techniques of controlling others and utilizing the powers of mind for selfish ends. This is something that's open to all of us. However karma never sleeps, and the more the evilists use their power for negative gain, the more imbalanced they become. Some of them, dwelling in the so-called "spirit" world on the astral level, have so much negative karma that their auras, or astral shells, are void of light and they cannot incarnate. They are able only to influence the physical level and those who dwell in physical bodies. To use mental techniques to influence the thinking of another ego is called "mentalism." Those individuals who *know* how much negative karma they have also realize they cannot achieve Mastership during this life-wave; thus they are out to bring as many egos down to their level as possible before God blinks and all non-Masters have memory wiped away. This notion accounts for the idea that God is "all-merciful" because the discreet bundle of mental energy is never destroyed—no matter how sick it is.

Since this is a learn-by-doing universe, it is through daily actions that we rise or fail. Nether influences try to get us to do those things that detract from soul

111

growth, and these evilists are far more clever than most of us. They pinpoint and use our weaknesses well. This fact gives rise to guilt complexes which compound the problems and we wind up with a confused world like that we see today.

Guilt is one of our major problems, but when one accepts Universal Law, he simultaneously realizes how silly it is to be guilt-ridden; karma exacts its dues whether we consciously reflect on our guilt or not, so we're better off not. Errors and misjudgments are common. The thing to do is forgive oneself and get on with the task of learning. After all, we've got centuries of negative conditioning to overcome, so forget the word "blame" and concentrate on obtaining knowledge. It's really simple, but we've made it so complex. For example, here I am with all this information and I still find myself entrenched in error and encountering problems regularly. Habits are hard to break, but we do have the opportunity to continue trying.

The sides were clearly drawn when Mu sank into the Pacific and the opposition heightened during the times of Atlantis and Rama. Finally we see in the priests of ancient Egypt and in the great teachers of ancient times, such as Moses, Buddha, Confucius, Lao Tse, Mahavira, and Zoroaster, the personification of the two forces within humanity.

The "righteous remnant" from Mu perpetuated their plan and stayed well within the rules of noninterference. The evilists, fortunately, aren't too terrific at teamwork, so nether influences have been disorganized for the most part, but on occasion they have been devastating: Witness the destruction of Ahknaton's efforts at enlightenment for all of Egypt; witness the strange way Adolph Hitler came to power and the resulting insanity that prevailed upon the world.

Now, is there any concrete evidence to show that any of what I've been saying is true? There certainly is;

112

in fact, it's one of the most concrete things on earth—the colossal structure known as the Great Pyramid in Egypt. This monument, more than any other, tells us of the Brotherhoods, of human perfection, of a Great Plan, and that the time is right for things to start happening in earnest.

If you haven't already read it, I'd like to recommend my book, *The Great Pyramid—Man's Monument to Man*. In that book the details are laid forth of how the "Elders" or Brothers managed to construct the Great Pyramid, beginning around 5,000 years before Christ and completing the job in 4699 B.C. to coincide with the "dawning of the age of Taurus."

Briefly, the Great Pyramid tells us so much in so many ways that it surely deserves the title "Man's Monument to Man." It is built of hewn limestone and granite, which were placed together with a precision unmatched anywhere on earth today. An estimated two and a half million blocks make up the bulk of the structure. The smallest of these blocks is limestone and weighs more than two tons. The largest blocks, huge beams of granite used to house the main chamber in the pyramid's geometric center, weigh an estimated eighty to ninety tons. No one has any idea how the task was accomplished. Esoteric information says the builders took plenty of time (300 years) and used canals and locks to transport the building materials to the structure's site. Only the tools and techniques that were generally available during the time period in which the builders lived were utilized. In other words, my information is that they did *not* have flying saucers from Atlantis and antigravity motors and the like as suggested by some metaphysical sources.

We know the isostatic adjustments of the earth's crust were pretty well completed at the time because the Pyramid builders oriented their structure to the cardinal points of the compass, and, after all these centuries, it still lines up accurately despite subsidence and earthquake.

The geometric perfection of the structure is unmatched. It was built on the design of the geometry of a sphere or circle. Each passage and each chamber is precisely and geometrically located, yet the construction is such that events in history are marked—in advance—clearly and without the problems of semantics because the markers are wordless.

Not only did the pyramid builders know more about geometry than was known to humankind at that period of time, but they knew as much or more about gravitational astronomy as we know today. They applied their knowledge of astronomy and geometry and built a massive edifice which embodies time and distance; which survived the ravages of nature and the plunder of man; which outlined the plan for the ages and prophesied historic events; which told a tale that points to the 20th century as being the "times of the judgment."

Those of us (English-speaking peoples) who use the inch inherited this handy little unit of measure from the ancient pyramid builders who evidently used it in Lemuria. The inch is the ideal length of measure since it is derived as a fraction of the size of the earth and is a perfect size for many everyday measurements. The ancient inch was derived from the polar axial length of our planet; someone measured the earth and determined the distance from the South Pole to the North Pole through the center of the planet—a consistent measurement for this spinning, plastic sphere. That distance was divided into 500 million equal parts to yield the inch.

The pyramid builders not only knew the perfect size of the planet, but in 5000 B.C. they knew that the earth orbited the sun once every 365.242 days in "apparent" time; that we orbited the sun once every 365.256 days in sidereal or star time; and that the orbit from perihelion to perihelion (farthest point from the sun) required exactly 365.259 days. They knew, and embodied the fact in the pyramid, that the cycle of *precession* takes precisely 25,856.25 years. Precession is the apparent movement seen in the heavens caused by the wobble of our planet as it spins and orbits. The stars appear to be in a different point in relation to fixed points on earth each year. The stars aren't really moving in the relationship; the earth is wiggling and causing the apparent movement. The annual rate of precession is very slight and it wasn't until the year 1905 that astronomer Simon Newcomb correctly computed the rate. The pyramid builders knew the annual rate of precession as well as the total period of precession.

This knowledge alone is enough to prove that our paradigm of history is erroneous. Let's take a long look at gravitational astronomy (simplified) and how such knowledge came to be.

Dr. Stelle and others have claimed that the Lemurian civilization lasted 52,000 years. Earlier I commented that this was significant because this time period allowed two complete cycles of precession to be completed during the period of civilized Lemuria. This means the astronomers of those times had absolute *knowledge* regarding precession. Our calculations are based upon less than five-hundred years of observation and are therefore nothing more than theory—albeit a sound theory since Newcomb's rate and the rate delineated by the pyramid match nicely.

Aside from the precise references to precession in the pyramid's dimensions, there are some obvious clues that the ancients had a fantastic knowledge of astronomy far earlier in time than can be explained by our present historic paradigm. To me the proof lies in the fact that such a thing as the "Zodiac" was known to all the early civilizations in Asia, the Mediterranean, and the Americas.

115

Of all the accounts of the history of astronomy I've read, none seem to stress this obvious point, but it is most pertinent. In order to have a "Zodiac" the observers on earth *must* be aware of the path of the earth's orbit around the sun. All the stars that make up the pictures used to delineate the various constellations of the Zodiac are located directly out from the orbital path. Sure, it would only take one year of observation to determine the earth's orbital path through the stars, but how many centuries of knowledge are needed to even conceive of such a thing? Our own history indicated a period of at least 4,000 years of confusion about the matter. If it is so simple, why did his contemporaries burn Bruno at the stake for arguing that the earth orbited the sun when the opposite theory was accepted at the time?

Knowledge of astronomy among "primitive" cultures has long puzzled anthropologists and archaeologists. The following paper, written by Stansbury Hagar and first published in *American Anthropologist* in 1912, exemplifies the academic approach to the topic and is still valid today. The article is reproduced in its entirety because the subject matter is pertinent to this entire book.

The study of the astronomy of the American Indians does not afford any definite evidence of their unity or diversity, or of the period or place of their origin. It neither proves nor disproves their origin in America or in Asia or in any other region. But it does present facts of value bearing upon the development of their culture, of relations between the peoples of America, and of their relations with the races of other continents in the prehistoric period of America.

Astronomy is not a primitive science. Long periods of years indeed must have elapsed before the really primitive man began even to observe the stars with anything less than utter ignorance and indifference, for they were related in no manner that he could apprehend with those material needs to which his attention was practically confined. Primitive astronomy began with the systematic observation of the stars to indicate direction upon night journeys, to indicate the hunting seasons to the hunter, and later to indicate the sowing and reaping seasons to the farmer. The cosmic and the religious element of astronomy— the questions as to the nature of the stars and their relation to the nature and life of man and of the cosmos—form, no doubt, one of the earliest bases of religious thought, if not the earliest, but such speculations, when they pass beyond mere wonder, surely imply a higher culture than the practical uses of stellar observation, and therefore cannot be earlier in time. It is evident, then, that the evolution of man must antedate the beginning of astronomy by a very long period of time. Even if we could trace astronomy back to its earliest source in time and place, it would afford us little or no information upon the origin of the earliest man, unless, indeed, he had remained in one spot during the whole of the long intervening period—a manifest absurdity.

Applying the above deductions to the American Indian, if he originated in America, astronomy cannot say when or where. We seem to see several foci of astronomical development in Peru, Mexico, and Yucatan, corresponding with general culture centers in regions having a climate and topography peculiarly favorable to the advancement of culture. The astronomical lore of all these regions is too nearly identical in complex concepts to be satisfactorily explained as due to similarities of race and of environment. There must have been an interchange of ideas between them either directly or through intervening nations in pre-Columbian times, hence we cannot be certain that this lore is indigenous to any one of the three regions named. Evidence of extensive migrations and of extensive change of climate in comparatively recent times adds to the uncertainty upon this point and prevents us from determining, at least in the present state of our knowledge, even the region

117

of the earliest astronomical development in America.

If the American Indian migrated into America from another continent in primitive times, astronomy would still be helpless to aid us in the search for the time and place of such migration because it cannot reach back to such an early period. But if this migration took place in later times or after the development of astronomical tradition, then indeed we may find in this field concepts sufficiently complex to render it possible for us to trace them back towards their birthplace. It is evident, however, that these concepts bear upon the origin of the American race only if they can be shown to be associated with the earliest race known to have existed on this continent—otherwise they will pertain merely to a later influx of an alien race into an already populated region. Pursuing this inquiry, then, let us ask first whether the concepts of American astronomy present such analogies with the astronomical concepts of other continents as to indicate intercommunication between them.

In the field of scientific astronomy the pole star was generally known throughout North America as the pivot of the sky, and the position of the South Pole was noted by the Peruvians. At least four of the planets were known and distinguished from the fixed stars by the Peruvians, Mexicans, Mayans, and some of the other tribes. The Peruvians had observed the sun spots, and a few among them were perhaps acquainted with the true cause of solar and lunar eclipses. All three peoples had divided the sky into true constellations and possessed a true solar zodiac. The Mexicans had ascertained the period of the apparent revolutions of foreign influence. The lunar and solar calendars of these three advanced nations, from the standpoint of the writer's cursory study of them, present little more evidence of intercommunication so far as their time periods are concerned, though the system of successive years governed by successive zodiacal signs recently discovered by Boll in Egypt and the Orient certainly suggests certain features of the Mayan and Mexican calendars. The presence in Peru,

Mexico, and various other parts of America of the Pleiades year of two seasons, divided by seed time and harvest, with its associated myths and rituals presents a stronger argument for intercommunication, one that has been elaborated by the late Robert Grant Haliburton, and Mrs. Zelia Nuttall has published evidence in favor of intercommunication based upon cosmogony and concepts which she believes to have been associated with the celestial North Pole.

When we enter the field of symbolic and traditional astronomy the evidence of intercommunication increases. We find among the common concepts the division of the cosmos among the four so-called elements: fire, earth, air, and water. The use of the swastika to express celestial revolution, of the cross and circle to represent the fourfold divisions of the sky and earth, of the serpent and egg with certain astronomical associations. Among the extra-zodiacal constellations, the Bear, formed by some of the stars of our Great Bear, has been generally recognized by the tribes of the northeastern portion of North America, probably from prehistoric times. It may be a legacy from the Northmen. The Milky Way as the Path of Souls of the northern tribes and the Celestial River farther south likewise finds European and Oriental analogies. But from the writer's standpoint the crux of the argument for intercommunication rests upon the symbols associated with the zodiac in Peru, Mexico, and Yucatan, for here we are considering not isolated analogies but an interrelated series in which the element of sequence affords an impressive guaranty against both chance and imaginative manipulation.

In Mexico the study of the elaborate system of judicial astrology may yield interesting results. So far as the writer is aware, little or no attention has yet been paid to this subject. In Peru evidence as to the zodiac is derived from the Star Chart of Salcamayhua, which names and pictures the signs, the monthly ritual which reproduces the attributes of the sign through which the sun is passing when the festival is held, and the celestial plan of the sacred city of Cuzco, which

119

was supposed to reproduce the observed design of the sky including the signs. This plan in varying aspects seems to have been typical of several and perhaps of many of the sacred cities or theogonic centers that form such a characteristic feature of American civilization. In Mexico the signs are named and pictured by Duran, Sahagun, Tezozomoc, in the Codices, and on the mural paintings of Mitla; their attributes are described in the monthly ritual and embodied in the plan of Teotihucan and in the day signs. In Yucatan the signs appear in the Codices, the ritual, the day signs, and the plan of Izamal.

As to possible European influence in these sources, the writer can only state his conviction that an examination of them will convince the student that such influence is either insignificant or totally absent. The following table will briefly indicate the correspondence between some of the concepts associated with the American zodiacal signs and with the signs we have received from the prehistoric Orient. It should be understood, however, that this table refers to only a few of the more obvious analogies:

SIGN	ENGLISH	PERUVIAN	MEXICAN	MAYA
Aries	Ram	Llama	Flayer	Stag
Taurus	Bull (originally Stag)	Stag	Stag or Deer	Two Generals
Gemini	Twins	Man & Woman	Twins	Cuttlefish
Cancer	Crab	Cuttlefish	Cuttlefish	Ocelot
Leo	Lion	Puma	Ocelot	Maize
Virgo	Virgin (Mother Goddess of Cereals	Maize Mother	Maize Mother	Mother
Libra	Scales (originally part of Scorpio)	Forks	Scorpion	Scorpion
Scorpio	Scorpion	Mummy	Scorpion	Scorpion
Sagittarius	Bowman	Arrows or Spears	Hunter and War God	Hunter and War God
Capricornus	Sea Goat	Beard	Bearded God	Water
Aquarius	Water Pourer	Water	Water	
Pisces	Fishes (and Knot)	Knot	Twisted Reeds	

120

Granting that these sequential analogies, if verified, establish intercontinental communication, we must now ask whether, if these concepts were brought into America from abroad, they seem to be associated with the earliest migration to this continent. We shall have to seek light on this point outside the field of astronomy. Professor Edward S. Morse and others have called attention to the significance of the facts that wheat was unknown in America at the time of its discovery by Columbus and that maize was then unknown outside of America; moreover, that there is little if any similarity between the more complex artifacts of America and of other continents. It is practically certain that the cultivation of these cereals and the manufacture of the higher grades of artifacts must have preceded the creation of a zodiac, and its transmission around the world, and it is not reasonable to suppose that a migrating race having knowledge of either cereal or of artifacts would have carried with them the knowledge of the zodiac without that of their food and tools. The inference is obvious. The knowledge of the zodiac was not brought to or taken from America by the earliest inhabitants of another continent, but must have been transmitted in later times.

We must still explain how such knowledge could have been transmitted in later times without the cereals and artifacts. There seems to be but one consistent answer. The transmission was accomplished by accidental or sporadic communication with individuals of an alien race who were able to impart their mental concepts but who brought with them few or no material products. There was no general migration at this time. Let the reader suppose himself unexpectedly thrown by shipwreck among a people with whom his race has never before communicated. Grant him a few companions only, and imagine the result. How much of their civilization would they be able to impart? Probably only a few ideas. They had no cereals and their attempts to introduce their artifacts eventually failed to overcome the force of conservative habit and custom opposed to change. This is admittedly theoretical, but it seems to be the only theory which reconciles the otherwise inconsistent facts. But if this explanation is correct, we see that even if the American Indian is a migrant from another continent, astronomy cannot help us to say when or whence he came, because as soon as we find astronomical concepts of sufficient complexity to afford a possible means of tracing them back to an alien home, they imply an advancement in culture inconsistent with the known characteristics of early American peoples, and therefore they cannot have come here with them. Astronomy reveals that there has been intercommunication with America in probably late prehistoric times, but it is silent as to what has taken place at an earlier stage.

While the author recognizes the evidence of similarities in astronomical ideas between cultures, the paper shows you how "science" is forced to react to anomalies, puzzle pieces that simply don't fit the picture. It also points out a perplexing problem even for the thesis of cultural traditions being passed down from Lemuria. It would seem that the ideas of agriculture and astronomy would certainly coincide as the author suggested, so how do we handle the problem of zodiac knowledge and obvious correlations between various depictions being passed along without corresponding knowledge of wheat in America or maize in Eurasia. Though it sounds facetious, why not consider that after cataclysmic disturbances occurred, the earliest peoples had the "knowledge" of the cereals, but they didn't have any seeds. If you and I were the only survivors of a cataclysm in a particular area, we would have all kinds of knowledge, but no technology or proof available.

The truly obvious connection in the zodiac (as I see it) is that all four cultures compared (which includes the Orient) listed the constellation Virgo as the "mother of agriculture" or cereals. The reason is equally obvious—Virgo occurs in the northern hemisphere during the harvest season.

Back to the builders of the Great Pyramid. Construction was completed, according to my theory, to coincide with the dawning of the age of Taurus. The date for completion was given in *The Ultimate Frontier* as being "about 4700 B.C." and no supporting evidence to authenticate the statement was provided. In researching the Great Pyramid, I learned from the works of Britisher David Davidson that the autumnal equinox (first day of fall) in the year 4699 B.C. coincided precisely with the star that makes up the toe of Castor in the constellation of Gemini. This clearly marks the boundary between Gemini and Taurus and, due to precession, our planet's axial plane was wobbling out of Gemini and into Taurus to begin the new "age." It should be explained that an age is an astrological term based upon some very sound astronomy. In a period of 25,856 years, during which we wobble through all the signs of the zodiac, the plane of earth's axis passes through each sign; twelve signs constitute approximately 2,100 years per age. However, the ancient Lemurians (this is conjecture on my part) used only six signs of the zodiac, each claiming precisely sixty degrees of space.

It can be shown that the early Chaldean astronomers and astrologers, who had the art refined to a degree second only to the Pyramid builders, used only the six signs, leaving us a clue that the influences of the balanced segment of Lemurian society (Brothers or Elders) also influenced the culture of the Tigris and Euphrates valleys. Later a compromise was made and we now have a mess as a result consisting of twelve signs of thirty degrees each that don't match the pictures. In addition, we ordinarily use day time reckoning (noon) rather than nocturnal reckoning (midnight) as the starting point for calculations.

I am convinced that this timing was part of the plan

that the Elders of Lemuria, carried along incarnation after incarnation, had laid out for progressively developing a future civilization commensurate with the heights achieved in ancient Mu. Incidently, this development would be no easy task; it took hundreds of thousands of years to get Lemuria started in the proper direction, and the plan was to realize the "Nation of God" in the 20th century, only 6,000 years after their plan opened at full throttle. The concept of "new ages" has been a firm one in our developing history. We opened with the dawning of the age of Taurus (4699 B.C.) and all over the Middle East, where the plan was implemented, we see people suddenly worshipping the bull. Now this was a far different kind of treatment for the animal we usually eat than what was afforded the creature in earlier times. The Minoan culture found in the Mediterranean and the early Ibers (Spain) had bullfights and bull sacrifices but didn't worship the beast in the same way the Egyptians and others did "Apis."

After Taurus came the compromise in signs and we find a ram or the lamb of God following the bull. Moses was harbinger of the new age and he really got frustrated when his followers tried to go backwards and worship the bull. Moses, according to *The Ultimate Frontier,* was a "High Adept," meaning he was almost a Master, and when he finished his task, which involved preparing the way for Christ to serve the plan among the descendants of the two tribes expelled from Mu, he did attain Mastership.

The advent of Christ ushered in still another new age—Pisces, symbolized in all Christendom with the sign of the fish. In *The Great Pyramid—Man's Monument to Man* I show how the birth and crucifixion are clearly marked in the Great Plan which the Brothers prophesied by marking the dates in stone for future generations to note.

Of course today we hear often about the "new age"

or the dawning of the age of Aquarius. The Great Pyramid also eliminates any confusion about the actual time of arrival of that age, for August, 1953, is noted within its structure, based upon exactly the same astronomy used to determine the age of Taurus; the link between ancient Mu and modern times is thus reaffirmed. Confusion among today's astrologers regarding the "new age" is due to the same old things—ignorance and free will. However, of all the art forms, astrology has held together better than most. Because today's astrologers use diurnal reckoning, they cannot accurately define the exact date for the Age of Aquarius which is based upon precession, the meridian of the Great Pyramid, and nocturnal reckoning.

Astronomers shake their heads in arrogant exasperation at this kind of talk. The astronomer who didn't read his history of astronomy very well will tell you, "There is no cusp, no dividing line between Pisces and Aquarius; so how can anyone tell when we precessed out of one age and into another?" The original sign, sixty degrees with very clear-cut cusps, was the sign of the fisherman. In August, 1953, we precessed out of the Piscean half and into the Aquarian half of that ancient constellation.

The Plan for the Ages, in retrospect, is quite clear. All the great teachers who preceded Christ by approximately seven-hundred years were part of the Brotherhoods. Guatama Buddha, Zoroaster, Lao Tse, Confucius, and Mahavira were all born within eighty years of one another, and all taught concepts designed to help the world's population understand the need and the responsibility of advancement. Christ couldn't have been more concise when he said, "You, too, can do as I do." He intended that everyone may attain human perfection.

Ignorance and free will almost botched Christ's message up totally, but the notions of freedom, love, humility, and virtue have managed to become firmly implanted despite ritualistic and interpretive distortions of the great teachings which have occurred in the meantime. But, as we look back through known history, we are able to chart the path of the Brotherhoods as they have worked to unobtrusively carry out the plan. Again, to illustrate, I quote from *The Ultimate Frontier*:

The century following Isaiah's advent was truly remarkable for the appearance of religious leaders along the whole of the Eastern trade route. All were high Adepts of the Brotherhoods, and each of these leaders was successful in reawakening religious understanding among the people to whom he revealed insights into the meaning and conduct of life. The impressive list includes Zoroaster the Iranian, Lao Tzu the Chinese, Mahivira the Jain, Buddha the Indian, and Confucius the Chinese. These are probably the greatest men to have ever appeared in the East; yet all were born within a span of 79 years—between 630 B. C. and 551 B. C. The impact of the teachings of these men upon their respective peoples was such that each inspired a religion which still flourishes. By 30 A.D. almost the whole known world had been exposed to great truths. However, the people who heard these teachers obviously were not ready for great truths, for the teachings of all of them were soon corrupted almost beyond recognition.

The Zoroastrian philosophy deteriorated into astrology and mysticism. The Jain and Buddhist ideals were diluted by the preposterous beliefs of the Hindus, so that now the Jains are mostly a food cult, and the Indian Buddhists believe in the desirability of no desires. Buddha, on the other hand, preached right action and right desire. He dreamed of brotherhood and tolerance but his ideas never dented the Hindu caste system. Lao Tzu's writings have inspired the Taoistic cult of magic and vain contemplation of obscure interpretations of his cryptic book. The Prince of Peace was called upon to sanctify the Crusades

126

and other wars—all of which are "holy," of course. Christ's sacraments have been appropriated by fee-hungry ministers; brotherhood is rigidly defeated by the congregations assembling in His name; and the beautiful, simple truth He taught has been interpreted out of existence by theologians. Only Confucianism has remained near its founder's intent.

The Brotherhoods take the human element into consideration and allow plenty of time for their plans, for they know how many years are needed to promote even one small philosophic change. A nation will often go off in the wrong direction according to its collective free will; so the Brotherhoods usually develop several reserve agencies to achieve their goals in case the primary agents fail or decline.

One of the Brotherhoods contrived to uplift the Grecian culture to the point of becoming an intellectual springboard for the Christ's deeper teachings, but the perverse character of the Greeks seemed to persist in remaining impractical, frivolous, and emotional. Because of their fratricidal wars and moral deviations, they did not benefit from the Brotherhood's program but fell into decline in the third century B.C. Nevertheless, their language *did* become a universal tongue for the spread of the Hebraic and Christian doctrines.

Alexander the Great was instrumental in establishing the Greek language as a universal tongue by his conquest of virtually the whole civilized world of his time—about 330 B.C. His overwhelming, relatively easy success was not merely a freak of good fortune; the accelerated exchange of learning which followed his consolidation of diverse cultures was much too valuable to Christ's impending arrival.

The Romans inherited the job of spreading Christianity through Europe by default of the Greeks. As a people, the Romans were very practical, emotionally austere, and enterprising in character. Therefore, the Roman Empire was at its height during Jesus' life. Christ came to uplift the people of the world, but we human beings were not yet ready to receive all He

127

had to offer us. As in the New Testament allusion (Hebrews 5:12-14), one must first learn to assimilate milk before one can have the meat. We certainly were barely ready for the milk.

During Christ's stay two thousand years ago, the world was a horrible place to live. Other than the very small upper class, there was nothing civilized about civilization. Even the patrician and king lived barbarously and brutally. In our vilest imaginings we cannot approach a true concept of life as it was in those days. The average life span was twenty-five years. Filth, disease, hunger, poverty, unspeakable oppression, hideous punishments for minor misdemeanors, and slavery at its worst were commonplace; and a strolling man or woman or child might commonly become a test for the range of an archer's bow or the soldier's new sword. There was no social legislation, and even children were exploited, maimed, and murdered with virtual impunity.

The establishment of the Roman Church through St. Paul paved the way to law and order in Europe. Before the era of the Church dawned, tribal lords, kings, and chieftains raided, slaughtered, and pillaged without end. With enforced obeisance to the Roman Church came the concept of protected, established, and defined feudal boundaries. Law enforcement eventually became feasible, and written laws were recorded by a learned, literate priesthood and the court scribes. The clerics who formulated the system often created kings and noble families to run it. The Church *was* government. More likely than not, bishops and cardinals actually ruled Europe while childish nobles were but pawns. The efficiency and stability of their systems were not effective overnight, but a certain degree of security entered the lives of European tribes for the first time. With security came establishment of an economic system which could be trusted from year to year; and as security grew, leisure could be bought with accumulated wealth. With leisure time came learning, with learning came the university, and thereafter came the rapid exchange of

128

ideas. The foremost task of the Roman Church was completed, and its authoritarian discipline was ready to be challenged. The political and economic security it established in Europe was then firm enough to sustain the Reformation. The resultant intellectual freedom during the Renaissance promoted the phenomenal advances of culture that we now enjoy. But we still stand *far* below our goal.

We shall soon enter the Aquarian Age—an era of great importance. The concepts Christ expounded two thousand years ago have been slowly taken to heart by many people. They have refined these concepts and have passed along their mental and moral advancement to their children and communities in an ascending spiral so that now a nominal proportion of the world practices natural attitudes of humaneness. At last Christ has some reasonable raw material with which to work—a far cry from the wretched creatures He found two thousand years ago.

Thus far we've come a long way toward establishing a new paradigm that accounts for more puzzle pieces than the present paradigm. Here's what we have so far:

1. There is a Creative Force—beyond our ability to know at this point, but obviously "there," that is holding the bag and laying down the tenets of universal law.
2. The planet earth was born, built up, and put together to best promote life based upon the chemistry of the carbon cycle.
3. The crust of the earth is subject to changes which violently affect life living on it.
4. Within that framework, a team of higher beings who have the same qualities of mind inherent in both the Creator and the created, but who are not yet so perfect as the Creator, yet far ahead of the newly created, set about to install discreet bundles

129

of mental energy into physical bodies so they could begin the arduous process of learning.

5. These higher beings worked hard to evolve a proper species to receive the gift of mind and finally developed *homo sapiens*. All the "tricks" of microbiology and the "intelligence" inherent in cells, genes, and so forth are under the jurisdiction of these higher beings.

6. The higher beings tried an experiment, allowing man to have a perfect environment so he could use his leisure time to seek knowledge. Instead, man chose to eat, drink, lie in the sun, and generally goof-off. The higher beings, being only human, bickered and finally without the full cooperation of all the higher beings, one of them pulled the plug on the edenic state.

7. Mankind had to struggle and learn to survive. In this way he learned to appreciate his fellow man and the benefits of "civilization." With the help of "borrowed" higher beings, mankind on earth developed a tremendously advanced civilization on a huge continent, Lemuria, now beneath the Pacific.

8. Change, which seems built into the universal system, finally caught up with that ancient civilization and cataclysm destroyed the continent. Man, having free will, but not having all truth, began to travel in tangents not particularly conducive to soul growth.

9. Civilizations of exceptional quality regarding technology developed in the Atlantic and India, but change also eliminated them. A remnant from man's earliest culture which has learned the truth about advancement and the purpose of life set out to help mankind regain the "golden age of Lemuria."

10. Sticking to the rules, which say one ego cannot interfere with another without express permission,

the Brotherhoods are slowly bringing civilization around to a point where it might be practicable to usher in the original idea.

11. The next step is to once again wipe the slate clean and hope that the remnant ends up a majority among the survivors of the cataclysm so the new civilization can be off on the right foot this time.

That's a summary of our new paradigm. Although tremendously oversimplified, that paradigm is far more workable and, frankly, more realistic than the present framework of understanding. The eleventh step is the one most relevant to us today. Since cataclysms have occurred in the past during highly civilized times, they most certainly could occur again in a similar setting. I have reason to believe that such a serious event is only a few short years away, and it's high time we gave the prospect some serious thought and made preparation.

If you're laughing that's fine, but remember: Though the Titanic was unsinkable, they still put a few life-boats on board.

CHAPTER 7

May 5, 2000: the Next Cataclysm?

Don't be deceived by my casual manner and relaxed terminology when dealing with this subject matter. I am firmly convinced of the veracity of the philosophy I'm sharing and have every reason to believe that another cataclysm will affect the crust of the earth on May 5, 2000—a scant twenty-three years from now.

In my mind all readers are skeptics, and that is how it should be. I adhere to the notion expressed by many teachers, especially the great Buddha: Do not automatically believe—have doubt. When we begin in doubt, we can end in knowledge. If we begin with belief, we end with faith—and blind faith is irresponsible. Faith, built upon rational fact, is justified. I have faith in the philosophy of the Brotherhoods and that predictions for a major cataclysm at the turn of the century are accurate. I don't *know* these things as fact because I'm not a Brother; however, I doubt that something so involved and interesting and complex and beautiful as

132

life and mankind can be purposeless. I also doubt that the Creative Force is the puppeteer over every ego and controls every single lifetime by pulling golden threads, or that all we have to do is *believe* to be "raised up forever." I believe in the responsibility for individual advancement belonging totally to each individual; and that we have thousands of lifetimes in which to learn the lesson of personal growth.

Now, let's get down to the potential for this coming cataclysm. Recalling what has already been mentioned about the size and location of the polar ice cap at Antarctica, it is plain to see that the potential certainly exists for the centrifugal effect to cause a crustal displacement. All that's needed, in theory, is a little something extra to trigger the effect. Year by year the ice has been growing heavier, and beneath the crust the pressure cooker has been building up momentum. The stage is set.

The trigger might be a particular alignment of the planets in our solar system around the sun; in fact it is a particular alignment occurring May 5, 2000 that makes that date predicted in *The Ultimate Frontier* credible. (Various prophecies indicating something calamitous is due with the turn of the coming century have been around for quite a long time; the most notable of the predictions comes from the famed French mystic, Nostradamus.) There is also a controversial planetary alignment projected for sometime in 1982, but if the grand conjunction of 1982 is the cataclysmic trigger, the various prophecies are incorrect, and academic stuffed-shirts can gloat that such prognosticating is less than scientific and that they told us so. If crust displacement occurs eighteen years ahead of the prediction I accept, I'll agree the experts were correct about prophecy, but not about cataclysm.

Before we get seriously into this notion of a heavenly perturbation acting as the trigger for the next cataclysm, let's consider what one of the wise old Egyptians had to say about such things from Plato's *Critias*:

> ... Thereupon one of the priests, who was of very great age, said: "Oh, Solon, Solon, you Hellenes are but children, and there is never an old man who is a Hellene." Solon, hearing this, said: "What do you mean?" ... "I mean to say," he replied, "that in mind you are all young; there is no old opinion handed down among you by ancient tradition, nor any science which is hoary with age. And I will tell you the reason for this: There have been and there will be again, *many destructions of mankind arising out of many causes. ...*"
>
> Now this has the form of a myth, but really signifies a *declination of the bodies moving around the earth* and in the heavens, and a great conflagration of things upon the earth recurring at long intervals of time. ... (Emphasis added.)

Earthquakes often make headlines in the news today, but in reality they are not occurring any more often than in centuries past—it's just that our electronic age records and distributes the corresponding information more efficently. I find it extremely interesting to compare the brief statement of the aged Egyptian above with the writings of two modern astronomers-seismologists, John Gribbin and Stephen Plagemenn, who coauthored *The Jupiter Effect*. The following is from the chapter entitled "The Planetary Alignment of 1982":

> Between 1977 and 1982 the planets of the solar system will be moving into an unusual alignment in which every planet is in conjuncture with every other planet; that is, all the planets will be aligned on the same side of the sun. Such an alignment occurs only once every 179 years, less than the period of Pluto's orbit (248 years). This occurs because the eight planets move faster than Pluto and so get round the sun ready for another alignment more quickly than

134

Pluto does. Neptune takes 165 years to get around the sun; so starting from a conjunction with Pluto, Neptune must complete one orbit plus another few years to catch up with the distance moved by Pluto in its orbit farther out. In the same way Uranus completes two and a bit of its eighty-four-year orbits while Neptune and Pluto are getting back into alignment, and while Saturn finishes nearly six orbits, Jupiter roughly fifteen, and the small inner planets whirl round at a giddy pace by the standards of Neptune and Pluto. This is why we can say that there are about five years during which the rare conjunction is building up. Each year from 1977 to 1982, as the earth moves around the sun, we will find the planets beyond Mars ever more accurately aligned. In the last couple of years first Mars and then the earth will move toward their positions in the alignment, followed by Venus. Last of all, little Mercury will spin round the sun completely four times during the year when all the other planets are lining up. Over a few critical months, there will be both a superopposition with Mercury on one side of the sun and every other planet on the other, and a superconjunction with all nine planets in line on the same side of the sun.

We have already seen how dramatic the effect on sunspots can be when similar alignments involving only Mercury, Venus, earth, and Jupiter occur. Is there any reason to believe that the superalignments will not produce even more dramatic effects? Certainly, dramatic effects have been expected from such auspicious events as long as man has studied the stars. Some astrologers mark the beginning of a new age by the occasion of the grand alignment—when Jupiter aligns with Mars and the moon is the Seventh House, the Age of Aquarius begins. The Age of Aquarius will be, we are told, a time of peace and love. But will it be ushered in by a major slip of the San Andreas fault and a wave of earthquake activity around the globe, unprecedented since seismology became a true science?

It is not only astrologers who are fascinated by the

135

alignment of the planets. Such a grand alignment provides a rare opportunity for space vehicles to be sent on a "grand tour" of two, three, or more of the outer planets. Current NASA plans are for two Mariner-class Saturn orbiters to be launched in 1985, followed the next year by two Mariner spacecraft that will use gravity assist to fly past Uranus and on to Neptune. It would be the literal truth to say that the launch dates of these space probes, which depend critically on the exact positions of the planets, are determined by NASA's expert team of latter-day astrologers!

So the next time of sunspot maximum will occur early in 1982, as Professor Wood has predicted. If the tidal influence of the planets neglected in Professor Wood's study is calculated, it turns out to be rather small. We might suppose, then, that the grand alignment of 1982 would be no more dramatic than other alignments involving Venus, earth, and Jupiter, although these are impressive enough. But a remarkable study published twenty years ago, in 1954, hints that this may not be the case.

After World War II, the study of radio communications entered a new era. It was known that the activity of the sun disturbed radio communications—we now know that this occurs through the sort of interaction of solar particles with the earth's magnetosphere discussed in chapters 7 and 8. Many radio engineers became interested in predicting "radio weather"; in other words, they wanted to predict solar activity so that they would know in advance when radio communications were likely to be difficult. Working in isolation from astronomers, the radio engineers tackled the problem in a completely empirical fashion. All they wanted was an effective way to predict the influence of the sun on radio signals on the earth; if their ideas conflict with established beliefs about the sun, just too bad. Well, they found such a "radio weather" predictor and it did run counter to some cherished astronomical beliefs, which may be why the results of these studies are not too widely known among astronomers. However, the prediction scheme was trusted by RCA Communications Inc., which financed the study of radio weather forecasting not through any scientific altruism but because of hardheaded business sense. Dr. John Nelson reported the results of his study in 1954, and in spite of their remarkable predictions, they seem to have laid dormant ever since, at least as far as astronomers are concerned.

Dr. Nelson and his team soon found evidence of the influence of planets on sunspots—the sort of evidence we have already discussed for Venus, earth, Mercury, and Jupiter. But moving rapidly on from this look at the solar system as a whole, they investigated relationships between planetary alignments and radio weather during the quiet period of the sun from 1951 to 1953, when there were very few sunspots. They found that alignments of 0 degrees (planets on the same side of the sun), 90 degrees and 270 degrees (when planets and the sun form a right-angled triangle), and 180 degrees (planets in line with the sun but on opposite sides) are significant indicators of the radio weather. When *any three* of the sun's nine planets are aligned like this, there are radio disturbances even when there are few sunspots (that is, during the years between sunspot peaks). More severe radio disturbances were related to alignments of five or six of the nine planets at these angles at the same time, to within a few days. The most impressive radio disturbances occurred when one of the inner planets (Mercury, Venus, earth or Mars) was linked in such a geometric arrangement with the sun and one or more slower moving planets (Jupiter, Saturn, Uranus, Neptune, or Pluto). From our point of view, one of the more important features of this discovery is that even tiny Pluto, on average between thirty and forty times as far from the sun as is the earth, and much smaller than our planet, plays a part in the disturbances by which some kind of activity on the sun affects the earth's ionosphere.

Sunspots, of course, are just the most visible part of solar activity. Cosmic rays from the sun are constantly streaming past the earth, and variations in cosmic rays will affect the ionosphere and radio propagation, even when the effects are far too small to change the atmospheric circulation, let alone change the earth's spin and produce earthquakes. Although Pluto itself is probably of very little importance in the chain from planets to solar activity to earthquakes, Nelson's remarkable work shows just how the whole solar system interacts to affect facets of our daily lives.

We are concerned only with the most dramatic disturbances of the sun's equilibrium, the sunspots, since to trigger earthquakes we need great disturbances of the ionosphere and the earth's atmospheric circulation (far more than is needed to disrupt the propagation of radio waves). We have guessed that it is the tidal influence of planets on the sun that is important, and we have seen how planetary alignments affect sunspots. Perhaps there are other factors also at work, as suggested by the radio weather studies of Dr. Nelson and his team of engineers at RCA. But of one thing we can be absolutely sure: the unusual planetary alignment is inexorably approaching and it will affect the activity of the sun. We have come a long way in our search for a trigger, from California to Pluto, but it looks as though we have found it.

What we have learned from Professor Wood's study is that the sunspot cycle now in progress is a long one of thirteen years. That is determined by straightforward calculation of the tides raised on the sun by the most important "tidal planets." So the cycle will peak in 1982; but we can go further than Professor Wood and say with some confidence that the activity of the sun around that peak year will be unusual even for a time of solar maximum. The reason for this is the whole series of unusually significant alignments of all the planets as they approach the "superconjunction" of 1982. As we saw from Nelson's work, even Pluto plays a minute part in affecting the sun at such a time; the most important effects,

138

however, are those of the tidal planets, especially massive Saturn, Neptune, and Uranus.

When Jupiter aligns with Mars, in the early months of 1982, the sun's activity will be at a peak; streams of charged particles will flow out past the planets, including the earth, and there will be a pronounced effect on the overall circulation and on the weather patterns.

Finally, the last link in the chain, movements of large masses of the atmosphere will agitate regions of geologic instability into life. There will be many earthquakes, large and small, around susceptible regions of the globe. And one region where one of the greatest fault systems lies today under a great strain, long overdue for a giant leap forward and just awaiting the necessary kick, is California. The situation is not directly comparable with that of 1809, the last time such a planetary alignment occurred, because we have no way of knowing how much strain the San Andreas fault was then under. The key to disaster is that this rare trigger should operate just when pressure along the fault is becoming intolerable.

Most likely, it will be the Los Angeles section of the fault to move this time. Possibly, it will be the San Francisco area which has a major quake. The prospect of both these sections of the fault moving at once hardly bears thinking about. In any case, a major earthquake will herald one of the greatest disasters of modern times.

Needless to say this book shook some people up and a rash of concern erupted across the country, especially among the metaphysical set who have so many psychic predictions of disaster on their ledgers, it's a wonder they can sleep nights. People were actually wondering if the world was going to end. The following quotation, taken from the "Bee Line" feature column appearing in the *Chicago Daily News,* illustrates the confusion aptly:

139

Q: Is it true that the world is going to end in 1982?—R.G., Summit, Ill.

A: Check back with us in 1983 and maybe we can give you a definite answer. What you apparently are referring to, however, is a unique alignment of the planets that astronomers say will occur at Christmastime of that year, the effects this will have on the sun and earth, and Biblical prophecies of what this may mean, as interpreted by at least some fundamentalists. The alignment ... will be one in which the planets are in perfect order on one side of the sun. It is only once in about 179 years that all nine planets are on one side of the sun at all, and perfect alignment has not been known in recorded history. And in the past, when the planets all have been on one side, things like these have happened. A great increase in magnetic activity on the sun, with huge storms, sunspots, and solar flares. Changes in prevailing wind directions. Radical alterations in the patterns of rainfall and temperature. An effect on the earth's rotation, changing the length of days. And the gravitational effects wrought by perfect alignment are expected to be far greater than those of the past, zigzag arrangements.

The dire prediction that this may mean the end of the world is based largely on these words of Jesus, from Luke 21:25-27, regarding his Second Coming (Judgment Day): "For then there will be signs in sun and moon and stars, and upon the earth dismay among nations, in perplexity at the roaring of the sea and the waves, men fainting from fear and the expectation of the things which are coming upon the world; for the powers of the Heavens will be shaken. And then shall they see the son of Man coming in a cloud with power and great glory." Also some astronomers fear that the solar storms that may be brought on by the alignment could trigger an atomic collapse of the sun. The Rev. John D. Jess, president of the Chapel of the Air Newsletter, Wheaton, spoke of 1982 in a radio broadcast on April 8, concluding with this warning: "Along with many of my Bible-studying colleagues, I believe we are very, very close to the end time. Now is the time for you to make your peace with God ... !" Also, the message was printed in the April newsletter, and a copy of this has been sent to you. ... But will the world really end in 1982? If it does, George Orwell certainly is going to be made to look like a bum prophet."

Although the newspaper editors handled the question with humor, the Bible quotations and the convictions of Rev. John Jess do make one really wonder what's going to happen at Christmastime, 1982. But then the giants of science appeared in the picture and with typical authoritarian arrogance, the Jupiter Effect was considered to be of no effect at all. *Science News* reported on the work of Gribbin and Plagemann in the December 13, 1975 issue:

The "Jupiter Effect," a theory of possible seismic activity on earth correlated with a 1982 alignment of planets in relation to the sun, has evoked little enthusiasm among geophysicists. The theory, set forth in a 1974 book by John Gribbin and Stephen Plagemann, poses a scenario in which the planets of the solar system, lined up on the same side of the sun, exert tidal forces on the sun to create an overabundance of sunspots. This increases the probability of a solar eruption, thereby increasing interactions with earth's atmosphere, which in turn affect the earth's rate of spin, finally triggering potentially severe earthquakes.

Sharp criticism has now come from astronomer Jean Meeus of the *Vereniging voor Sterrenkunde* in Belgium. First of all, writes Meeus in *Icarus* (26:257), the four major planets—Jupiter, Saturn, Uranus, and Neptune—will not be aligned in 1982 but will span an arc of some 60 degrees. Furthermore, he says, the combined planetary tidal bulge on the sun will be 2.7 million times smaller than the tides raised on the earth by the moon, and the evidence cited by the authors for a planet-sunspot correlation shows no such correlation. There are no proven connections between solar activity and the earth's rate of spin, he says, and the low correlation in a study of 21,873 earthquakes between 1910 and 1945 by J.M. Van Gils led to the conclusion that "seismicity and solar activity are thus mutually independent."

The authors, replying in the same issue of *Icarus*, agree that "the exact tidal mechanism postulated by K.D. Wood, (the one cited in their account) looks less plausible than it did when we wrote our book." They maintain, however, that other mechanisms could be substituted for that and other steps in their scenario without invalidating the ultimate result.

Meeus, in a terse counterreply, simply cites brief additional data to the effect that neither planetary influences on solar activity nor solar influences on terrestrial seismicity are established. "The Jupiter Effect," he says, "does not exist."

With that peremptory windup, the academic side of the controversy evidently closed and the two authors lost. However it remains to be seen if they are terribly in error. We just might experience some horrendous earthquakes and coincidental sunspots in 1982, or, on the other hand, we may have a fizzle of the magnitude of that Kahoutek Comet caper in 1975.

With the help of a postgraduate astronomer from Northwestern University, Harry J. Augensen, I am able to list the positions of the nine planets relative to the sun at Christmastime, 1982. It appears the Belgian is correct in his statements about the alignment. Putting the sun in the center of our diagram (see photo section), the planets rest on these degrees of a circle on Christmas Day, 1982: Mercury, 334°; Venus, 282°; earth, 94°; Mars, 337°; Jupiter, 236°; Saturn, 204°; Uranus, 244°; Neptune, 268°; Pluto, 201°.

Now, a look at May 5, 2000 (see diagram in photo section). Earth is all by itself on one side of the sun, a full 180 degrees off from nearly all the other planets. Heading off in a straight line directly into the constellation Taurus are the following planets and the degree line of the circle with the sun at center: Mercury, 42°; Venus, 92°; Mars, 51°; Jupiter, 42°; Saturn, 42°. Uranus and Neptune are triune, or off at 318° and 304° respectively, and Pluto is on the far side of the sun with earth. Earth is located at 224° and Pluto will be at 232°. According to certain esoteric information, this alignment will indeed trigger the cataclysm. It is notable that the moon will also be in Taurus at that time, meaning there will be an alignment effect consisting of either a push or a pull, or some kind of magnetic, electrical, mysterious thing away from the earth in practically a straight line due to the locations of the moon, sun, Mercury, Mars, Jupiter, and Saturn. Venus is off-line some 40 degrees, and Uranus and Neptune are off in left field.

The bulge of the ice cap at Antarctica is greatest at the 96th meridian, so it stands to reason the thrust will be out along that line. This means the crust will be shoved northward along the meridian that runs through much of the Indian Ocean before contacting a major city, Rangoon, Burma, then the rest of Asia.

Looking at a globe you can see that such a displacement must shift some plates toward the equator and some away from the equator. The bottom of the Indian Ocean will be stretched as the violent thrust northward drags part of what is now called the Australian plate over our planet's equatorial bulge. Far to the north, Siberia will be contracted and compressed and shoved even farther into the chill arctic region. An ice cap will immediately begin to form on the land masses of Siberia, except at such points where that continental plate may submerge beneath ocean waters.

Along the opposite side, we find that our meridian of maximum displacement runs right through Cleveland, Ohio, meaning that North America will move southward toward the equator, being stretched out as well as sloshed by ocean waters. Two segments of the earth's

143

surface, diametrically opposite one another, will be pivotal points and will move the least. One is located in the mid-Pacific near Hawaii if the assumption about the 96th meridian is correct, and the other is near the west coast of Europe. Such pivotal points are evidently less subject to serious damage—at least this is the theory of some cataclysmologists. It has been argued that the Malayan Peninsula of Southeastern Asia was a pivotal point during the great Lemurian cataclysm, and the flora and fauna of that area indicate little change over many thousands of centuries because of the relatively minor damage that was sustained.

It is part of my belief that Mu will pop back up out of the Pacific, and the next great civilization will inhabit and develop that newly resurfaced continent. This is purely philosophical conjecture, and I am at a loss to provide sound reasoning to support a contention that the Pacific plate should reemerge, but a geophysicist open-minded enough to entertain such a theory just might come up with something convincing in the way of evidence.

As of today our future looks pretty exciting, though some may object that it looks scary instead. The question is: What can we do about such an impending cataclysm and its devastating effects as of today? Is it really possible to rebuild a more perfect world from the demise of this one? Is there any way we can avoid such devastation? What's the best course of action?

CHAPTER 8

The Individual and His Incarnate Future

In the first chapter I pointed out how we people live our everyday lives on the basis of what we believe. At this point it is important that you understand that by writing this book I am not trying to make anyone believe what I believe—I'm not a *katholi*. You are responsible for your own beliefs, and, as such therefore, the purpose of this book has been merely to make information readily available to you. I think the information presented herein is reliable and accurate, but only you can decide on a personal course of action.

We are living in a most interesting period of time, which many believe coincides with the Biblical "times of the end." I agree with that belief, but my course of action has been different than that of most Christians. Perhaps it is testimonial to the veracity of Bible prophecy that one can draw similar conclusions without ever having been influenced by organized religion. Also, this same lack of church influence has given me a different perspective on these times; I have totally different definitions of certain oft-quoted words such as "rapture," "repent," "Doom's Day," and "Kingdom of God." I am convinced that my view of these concepts is rational, and I am certain that because of my views fundamentalists of Christianity are convinced that I am "of the devil."

I look at these important words not so much in a negative as a positive vein. *Repent* to me means to quit looking around for someone else to take responsibility for your advancement and take some positive steps on your own. Mean, nasty people like Hell's Angels motorcycle members are not the only ones that need repent; nice, pious, church-going people also need to repent—not so much for overt nastiness, but for simply figuring that Jesus is going to do everything for them if they *believe* and sing hymns once a week. Of course this is vastly oversimplified—there are many millions of people who attend churches and strive to live a life of virtue and knowledge. These individuals are the fodder for the Brotherhoods' Plan, for they are neither sinners nor are they religious fanatics.

Doom's Day is that event which follows the terrible war which, over the years, has come to be called "Armageddon." Many believe Doom's Day will be embellished with fire and storms and lightning and volcanoes and earthquakes and you name it in the way of disaster. They think it will have the effect of striking down the millions of evil people who haven't already had their brains disassembled by atomic warfare. Doom's Day is going to solve all man's problems. While the unpious, the irreligious, and those who don't pray regularly and fervently are destroyed, those who are somehow fervently "with Jesus" will be lifted up in a thing called the "rapture," saved forever, and ensconced in the kingdom of God.

I have certainly oversimplified this belief that is so very prevalent among Christians, but in essence the representation is correct. I don't buy it! If this version is true, then my mother was right—God is a strange character! However, it seems to me intuitively that the sequence of events may be quite accurate and that the chief problem is in the interpretation of some of the words which separate rationality from religious belief. Over the past few centuries theologians have interpreted a great deal of rationality out of the scriptures whether intentionally or not. For example, it is totally inconsistent to believe that the Creative Force is preparing to suddenly intervene and "lift" people to safety when the next cataclysm occurs, simply because they are huddled together, fervently praying in a building allegedly dedicated to Him.

Having been a student of phenomena, particularly healing phenomena for the past ten years, I am convinced of the power of human thought. I refer particularly to especially directed, affirmative human thought such as that we find in prayer. Even *knowing* this power exists, I'm betting that when Cleveland starts heading toward Mexico City on May 5, 2000, many millions of fervent believers holding hands in their temples and praying with all their might will be clobbered along with most others in the paths of destruction. Those who are somehow saved from the upheaval will maintain it was through prayer. I won't be able to argue with that rationale, but what will be the sensible explanation for the praying millions who don't survive? God only knows!

I define Doom's Day as the name for the cataclysm that is due on May 5, 2000. If it comes early or late it will still be Doom's Day. As I see it the events coming up will approximate the following description: First, there will be economic chaos which will be a blessing in disguise for America. No matter its method of appearance, such an occurrence will serve to stop us from trying to protect and finance the rest of the world at an expense we obviously cannot afford and may even

147

break up the greatest form of central government since those of Lemuria and Atlantis. Rather than disassemble those constitutional principles so vital to our civilization, economic disaster will primarily serve to break up a huge, cumbersome bureaucracy that has begun to do more harm than good. The tools of Armageddon were forged in Chicago beneath the bleachers at Stagg Field and were tested on the Japanese at the end of World War II. Politics are already rapidly dividing the world into armed camps for the most violent war yet endured on our planet. Near the turn of the century the buttons will be pushed and atomic weapons will be electronically unleashed . . . and Armageddon. I believe the economic collapse will happen because we do not have a dictatorship and because we will pay economic karma before the mid-1990s, but Americans will not be generally involved in this terrible atomic war. After atomic war wrecks the environment, many struggling survivors will consider cataclysm a blessing.

Such an upheaval just may be the only efficient way to clean up the planet after such insanity. Doom's Day, however, is not the end of the life wave; it is a gigantic housecleaning, and those who survive with technology and knowledge intact will be able to forge a new civilization based upon the principles that have made America the greatest nation since Atlantis, and perhaps on the principles of ancient Lemuria.

Now, how do we manage to survive with technology as well as our skins intact? *The Ultimate Frontier* offers a far-out, but challenging and possible solution: Aircraft capable of lifting thousands of people into the stratosphere and returning them safely when the worst is over. They will land with the purpose of rebuilding civilization on more positive principles. They will begin building the Nation of God.

Yes, I know it sounds utterly fantastic, but it is far more rational than the other Christian version which I personally assume is meant by "leaving in the rapture."

The number of people to make this excursion has been prophetically set at 144,000. And, I have been told in private conversation, this stems from the wish of the planners to have 12,000 souls from each of the original 12 tribes of Mu. Regardless how many people are elevated or what their original tribal status was, it's curious how such an airlift could be construed as a "rapture." Perhaps in his vision John the Divine saw the thousands of airlifted citizens in a state of suspended animation, induced to help them make the two- or three-week flight without too many problems.

Can you imagine all those people jammed into aircraft under "normal" conditions? Think of it—crying babies, snippy teen-agers, love affairs, family spats, and worry worts all tossed together while their world crumbles below them. Indeed, the riders of these craft will need be patient, virtuous individuals of unmatched courage and determination. The prospect of an unstable, undetermined future alone could cause the strongest nerves to crumble during the flight, which could last up to three weeks. In fact, it would seem to be far easier to merely let the cataclysm clobber you and then reincarnate later into a better world.

Another view of the rapture is totally different than what I'm considering here, and that is the notion of what it will be like at the changing of the Life Wave—that time to come when God blinks and everything changes and those who have made Mastership step up to the next level, and those who didn't have their memory erased. This may more accurately fit many Christian views of the rapture than rising up from the holocaust. Either way, it's pure belief.

I've been told that there are literally millions of advanced egos hanging around in the ethers waiting to incarnate into a civilization—that is, a *positive* civilization where citizens won't be conditioned as tiny children to lose whatever advancement they may have attained during previous lifetimes. If that sounds unrealistic to you, think of the millions of children incarnating into our society today who will have their beliefs shaped by today's environment!

Is such an airborne plan feasible? This is the kind of challenge science fiction writers love—working out the logistics of building aircraft capable of carrying hundreds of people, farm animals, seeds, equipment, and other necessities. Airships that will comfortably attain an altitude of about fourteen miles and "hover" or cruise for more than two weeks, then descend and land safely on unprepared terrain. If the figure of 144,000 is accurate, then we'll need a lot of aircraft; so who's going to pay for them? Assuming they can be financed and built, what mode of power can we use? Of course the bureaucrats will demand an environmental impact study first.

Agreed, at this time it's an improbable task. But, there are people busy working on a solution to the logistics of this particular "airlift" at this very moment. Those of us who belong to *The Stelle Group*, which is the organization that publishes *The Ultimate Frontier*, believe that throughout the world there are several groups of individuals who are preparing for the prophesied events of the next twenty-three years. In many ways *The Stelle Group* is unique, but we do not claim to be the "only" way. *The Stelle Group* started as two people, Richard and Gail Kieninger, in 1960 and today we have a small but elegant community with fine homes and a bustling factory located in a cornfield near Kankakee, Illinois. We are working to accomplish two primary goals: First we are creating an environment that is conducive to individual advancement based upon individual effort, especially with respect to our children. Secondly, our goal is to bridge the gap that will be left by the prospects of destruction ahead. We intend to survive, not only with our skins, but with technology, so that mankind need not revert again to a stone age existence. We intend to implement the great plan of the Brotherhoods as best we can. Even if the second part of our project is not necessary, and we'd dearly love to learn that all the prophecy of economic disaster, Armaggeddon, and Doom's Day is mere fantasy, we will have a magnificent city peopled by responsible citizens.

We do not feel we are alone. I'm convinced there are many groups and individuals taking part in this program, and simply because I cannot name them or locate them doesn't mean they don't exist. No single person or group could have such an awesome responsibility. There are literally millions of people available to help implement the plan.

Part of my father's credo that rubbed off on me was to be very wary of anyone or any group that promised perfection and salvation and so forth. Despite my conditioning from that irrepressible cynic, I've joined an organization—but look at what has been promised: hard work, pain, doubt, difficulty, and hope. After eight years I'm still quite steadfast in my commitment.

Many of you will want to investigate *The Stelle Group;* please do, but don't expect any of us to try to convert you—we do not proselytize. However, we do want and need members and all the help we can get. Not since the building of the Great Pyramid have the Brothers needed so much support for a single task. Not only must our community grow and prosper to enhance the environment as it should, but if we are going to build the aircraft necessary to overcome the effects of the cataclysm, we'd better get started.

It is realistic to set a goal of building aircraft that can carry about 200 to 500 persons because our present technology has accomplished this feat. We will require some kind of power units that do not rely on fossil fuel. Sounds a little like we need to build a flying saucer type vehicle, doesn't it? That's it! We need to construct a fleet of IFOs—we certainly can't call them UFOs because if we build them, we can identify them.

151

This is not fantasy, it is *seeing the possible,* which is a trait sorely needed today. However, before I launch into considerations of the controversial topic of today's UFO phenomenon to see if any of it applies to our project, I need to make a pertinent observation about a social trait prevalent today. It is perturbing to many of us in *The Stelle Group* that the news media insist upon emphasizing the "Noah's Ark" theme and UFO phenomenon which simply are not part of our thinking, rather than stressing the uplifting aspects of our community. Therefore, when I start talking about UFOs many people immediately jump to erroneous conclusions. Indeed it is a lot like the tale of Noah and his Ark that we are considering an airlift necessary—but our airships will not go out into space, nor are we "believers" in the various UFO notions circulating today.

Something has been sighted that doesn't fit any known paradigms; however, the honest-to-goodness sightings have been extremely rare. It's easiest to say "there's no absolute proof" of flying saucers and let it go at that. Of all the phenomena I've investigated since 1963, the so-called UFO phenomenon is the least impressive. Sighting an object or point of light which appears to make unique movement doesn't give researchers much to work with.

I am reminded of the standard discussion of visual perception employed by many Buddhist teachers whenever a person tells me: "I saw something last night." The Buddhist Lama explains that if a person sees a tiny dot on the distant horizon, his vision observes a tiny dot on the distant horizon; but his conditioned mentality will construe the object to be either a tree or a rider. He cannot see green leaves shimmering in the breeze; he cannot determine from that distance the nature of any beast of burden. He sees a dot on the horizon and his thinking, his own version of reality, makes what he sees more than just a dot. However, af-

152

ter taking more time and moving closer, he can visually confirm or deny his tentative conclusion.

In my opinion, and in the opinion of many researchers, this is precisely what most UFO observers are doing—seeing something and simultaneously painting a mental picture of what they want that something to be. However, some observers have gone well beyond that. Take the example of Peter Reich, a former coworker of mine at the now extinct *Chicago Today* newspaper. He is an award-winning aerospace writer and one of the original Whiz Kids on radio. He had no use for UFO phenomena and was a technocrat of the first order. One night from his apartment window, Pete saw a bright object. He watched it, calculated its movements with rough triangulation using his finger against the glass, carefully observed the object for several minutes, then concluded it could not have been a flying machine known to our technology—and who should know better than a flying machine expert?

Others have seen vehicles up close; at least that is the claim. Some people have allegedly been taken for a ride in these spacecraft. Most of us are familiar with those stories. I remain skeptical of these so-called "contactee" phenomena because it is not impressive to me that some of these people can pass a lie detector test and even tell a straight story under the influence of sodium pentathol, etc. It is a simple matter for nonphysical or nonreal experiences to be impressed upon the mind of an individual so that he is absolutely convinced of the reality of the experience. Hypnosis is something familiar to everyone; mentalism is a form of hypnosis that may be harder for most to believe than the prospects of UFO trips being real events. However, the evidence seems to indicate mentalism rather than actual physical contact.

I am relegating the UFO contactee to the level of a "spiritual dupe" or victim of influences from mentalities belonging to discarnates. Look at it this way: If a

153

discarnate intelligence can obsess a Brazilian peasant so that the uneducated peasant can perform utterly fantastic feats of healing using absolutely unbelievable methods, then a discarnate personality can most certainly make some individuals believe they are taking a trip in a flying saucer. The obsessing phenomenon I am referring to was demonstrated by one Jose Arigo whose story has been well told by John G. Fuller in his book, *Arigo, Surgeon of the Rusty Knife.* I did not observe Arigo firsthand, but I have observed the equally fantastic Filipino psychic surgeons firsthand on many occasions during several years and authored the book, *Psychic Surgery,* in which this type of mentalism across the planes of existence is thoroughly demonstrated.

In that book, I used the illustration of "telepathic hypnosis," alleged to be proven by scientific observations of the famed Hindu rope trick. In brief, observers of the rope trick saw the fakir perform incredible feats which defy all physical understanding, but when films of such feats were developed and shown, the fakir was doing nothing while the ring of observers around him were obviously being impressed as evidenced by the looks of awe and bewilderment on their faces as their eyes followed the nonexistent action.

In the same vein, the people who have experienced UFO contact are not lying—the experience was genuine to them. I could make some wild guesses about the motivation of such intelligences for impressing such experiences on people, but they would not be particularly germane to the discussion. For those who doubt ethereal existence and are turned off by any mention of "spirits" and the like, let me stress that phenomena is not a figment of anyone's imagination—it is just as "real" as a chair or a table. The accepted theory of matter tells us that the chair and the table are, in "reality," whirling electrons, protons, and neutrons so arranged as to make up atoms, then molecules, then hard wood furniture. To say that ethereal "things" are un-

154

real is every bit as ridiculous as the opposite view held by some Oriental concepts.

While you and I tend to maintain there is no reality except that which manifests itself physically, the Zen Buddhist stresses that there is no physical reality; that the physical plane is merely agreed-upon illusion. They feel that everything is vibration, or more correctly "the vicissitudes of yin and yang." In essence, perhaps this is correct—I think it is. However, there *is* a physical plane because we can sense it in five ways and agree upon its characteristics among ourselves. The yin and the yang have somehow been ordered by space and time into atoms of the elements from which physical objects are constructed. That physical evidence, such as chairs and tables, exists whether the Zen Buddhists agree with its presence or not. It is true that matter may be "materialized" by mental means if a person knows the techniques, but it is not true that matter on the physical plane is merely an illusion.

We've swung full circle and we're now back to the reality of UFO phenomena. So far, to my knowledge, the only physical evidence of UFO phenomena has been unexplained spots on the ground and something called "angel hair," neither of which tell us much more than "something has been there."

One final comment on UFO contactees: If these superior beings are hovering around waiting to "help" mankind, why are they handling such a supposedly noble task with such apparent stupidity? It is against the rules of "self-responsibility" for some superior intelligence to fly obtrusively into our environment in order to "take over" for our own good. The higher beings who are helping are not going to interfere so blatantly and so stupidly.

It might, I suppose, be argued that the Brothers are in many ways like the UFO creatures I dismiss; they are slipping around trying to uplift mankind without getting caught and so forth. It is definitely not the same. Whereas we have UFO contactees announcing that on a certain date a group of space vehicles will land and a pack of wizards will then come forth to solve all our problems, the Brothers constructed a marvel, the Great Pyramid, to evidence their plan; they have worked unobtrusively to help people help themselves. It is not reasonable to ask that the Brotherhoods function openly; they would not have a chance. If a Brotherhood tried to openly set up shop in America today, using everything afforded by their advanced capability, they would be feared and legislated against. After all, look what we did to Christ!

You can imagine, on the other hand, the tumult that would result if a spaceship were to land unannounced. Our government would be so petrified and suspicious that the occupants of such a craft would be napalmed or similarly attacked as they emerged. If they landed in America, the Communists would be blamed, and if they landed in China or Russia, America would get the blame.

Much of the information that has come my way is what I call third-class information. This category contains what may or may not be true and needn't be proved one way or the other at the present time. For example, the Lucifer-Jehovah story I related earlier could be true, but the story doesn't need acceptance or rejection at this stage of my growth. One tidbit of third-class information deals with the UFO problem. I've been told that the "saucer-shaped" vehicles are made and operated by members of the Brotherhoods. They use them and do their best to keep out of sight but are occasionally spotted or picked up on radar. The Brothers do not want to so blatantly interfere with our environment as would obviously happen if such craft were captured. Can you imagine the bureaucrats of the Pentagon or the Politburo reacting to the discovery of a genuine flying saucer? In the same exchange, I was told that the cigar-shaped UFO vehicles are actually observers from other planets peering at us much as we observe the moon and Mars. They are advanced enough to know the rules about interference. Well, it's just information—good for a chuckle if nothing else.

Here we are with a potential cataclysm staring us in the face and a need to build something that works the way we imagine flying saucers work. What was it someone once said about necessity being the mother of invention? One of these days the necessity will dawn on several million people, talent will come together, and the engineering problems will be solved. In fact, I believe the beginnings of the solution are popping up all over the place as one rugged individual after another announces the discovery of "a better way."

For example, Ed Gray is an uneducated Californian who has put together a remarkable power plant using electricity in a way that no other person has used it. Though his invention appears to be a financial loser, thanks to suppression by present-day attitudes and our economic system, he has succeeded in opening the doors of a new technology. There is also the record of remarkable experimentation and demonstrations of how to generate power from the "cosmos" put on by Dr. Henry Thomas Moray of Salt Lake City in the thirties. Moray actually generated usable electric power by hooking up a simple antenna and a series of solid state circuits (he used germanium transistors thirty years in advance of their invention by Bell Laboratory scientists) with a special cathode ray tube. His source for the energy was simply "radiant."

These are just two of the many innovations that are the potential means for developing proper technology to airlift thousands above the impending holocaust. I have no doubt that we can develop the necessary technology, without the aid of any UFO occupants.

Once the aircraft are built, tested, and ready for use, who gets to go and who must stay? That question appears to have all the ingredients for a terrific movie plot. A bunch of "kooks" figure the world is coming to an end, so they work hard and despite laughter from others, they build these flying ships. Then, when it's obvious they had the right idea all along, the hordes of skeptics descend on the ships in panic—everyone wants to leave in the rapture; in a frenzy. You design your own ending to the movie.

Since most individuals who adhere to this concept of cataclysms don't really fear death, there will be no panic. The selection of passengers will be orderly and motivated by what we have come to call "the highest good for all concerned." This means there will be some excellent pilots put aboard first, then plenty of healthy, intelligent young people who willingly accept responsibility. The details have yet to be worked out, but I am told there will be representatives of every race of man taking part in this noble experiment. I have been told also that the cataclysm is not going to "purify" the races which have been cause for so many current social ills. Actually, the races have not been the *cause* of anything. Instead racial problems stem from *attitude,* and our attitudes are designed by our beliefs and our conditioning, and once again we are face-to-face with ignorance and free will. Part of egoic growth, one would think, is the ability to genuinely overcome the superficial problems that can develop either directly or indirectly as a result of "race." I imagine that's why the higher beings evolved different subspecies in the first place to give us the opportunity to work out superficialities and get down to the nitty-gritty of human potential.

This presentation has been cursory and perhaps flippant in many respects, but as I stressed earlier, we are very much in earnest. It is part of my philosophy to maintain a sense of humor to avoid breeding anxiety. A good sense of humor is vital to the well-being of man and the concurrent growth of character. We

158

should even be able to manage a smile when contemplating something as devastating as cataclysmic change and the "times of the end." It isn't truly going to be the *end*; it is a new beginning just as every day is a new beginning, every birth is a new beginning, and every death is a new beginning.

What the new civilization will eventually be like depends on how well the forthcoming "remnant" does its job. My prejudice may be apparent by the inclusion of the following remarks written as an epilogue to *The Ultimate Frontier,* but so be it. According to the biographee, Richard Kieninger:

Persons who are predisposed toward formalized philosophy might have preferred the tenets of the Brotherhoods' belief to have been developed step-by-step from *a priori* foundations, but to undertake a logical argument on this basis would be unprofitable since philosophers are not yet in agreement as to what constitutes a truth firm enough to be regarded as a basic philosophic foundation. For example, mathematicians and physicists agree to ignore the many unprovable foundations of science so long as these arbitrary assumptions continue to work in practical applications. Even empiricists are often forced to accept as fact many things which are but subjectively self-evident.

In any event, it is highly improbable that the cosmology known to the Brotherhoods could ever have been imagined by man—let alone proven logically by him. The realm of scientific philosophy lacks objective evidence of the planes of existence beyond the physical, and without acceptable observations, an hypothesis cannot be derived nor understanding be achieved. Unfortunately, scientists will never acquire objective knowledge of the other planes of existence because the necessary evidence can only be obtained through mental experiences which are wholly subjective. Inasmuch as observations must be reproducible under laboratory conditions in order to be admissible

159

to the body of scientific knowledge, the powers of mind are officially cast into limbo. Thus we are faced with a crippling shortcoming in man's search for truth. Science has by its own rules limited itself to material phenomena and excluded itself from analyzing man's place in the cosmos. I don't quarrel with this sensible limitation, but I find it unfortunate that scientists have not as yet been able to extend their thoughtful probings to a disciplined study of the phenomena of the higher planes of existence. Essentially this has been the work of the Brotherhoods; and when a man sincerely undertakes the quest for higher understanding, he attracts the assistance of the Brothers who traveled that same path before him.

Our knowledge about the world has its root in man's perception of existence through his five basic senses. Observation and logical inference have thereby brought him into possession of a great deal of knowledge concerning his physical environment. The next step is taken when an individual has intensified his senses beyond the physical, for from that point onward his increased perceptions of nature allow him to know things about his environment that are closed to normal men. The ego who has advanced to the point where he is acceptable for admission into the Brotherhoods *knows* that the Brotherhoods' philosophy is truth because by then he is able to rely on empirical evidence—subjective though it may be. Men who have developed clairvoyant abilities are generally unable to convince those who haven't yet come to possess these powers that there are other planes of existence. It is like describing a rainbow to a man totally blind since birth—for all the blind man knows, the very idea of sight is a taunting myth.

A hypothetical island wherein all the inhabitants have for untold ages been genetically sightless would very likely declare a visitor with normal sight insane were he to describe the beauties of the sunset and the starlit skies. These islanders would have no place in their philosophy for a fifth sense, and their science would be seriously hampered and distorted by lack of the very important sources of information sight affords. The true clairvoyant fares no better in our society when he describes the beauties and wonders of the astral plane. He becomes the object of ridicule and stands in real jeopardy of being committed to an asylum if he has the temerity to insist. The intelligent, scientifically oriented person who has experienced spontaneous flashes of clairvoyance is predisposed to regard these experiences as mere coincidence or psychological weakness, and he consequently suppresses his emerging latent powers more vigorously. As a result we find the type of individual who can best forward mankind's quest for civilization—the critical, discerning, and logical ego—is turned from a true understanding of himself because of the code of thought prescribed by scientific dogma.

Fortunately, the current popularity of psychology mitigates an otherwise gloomy prospect for mankind's future. Even though psychology is not yet a science (being essentially still in the stage of amassing observations), many of the reasonable conclusions it has synthesized after a century of analysis have earned it respect in scientific circles. The psychologist's evident ability to predict human behavior has led to several accepted "laws" which are of practical use. Parapsychology is likewise beginning to draw attention to the case for man's extrasensory perceptions, and sheer weight of evidence may eventually force the scientific philosophies to accord serious consideration to the sixth sense.

Social psychology, led by the depth psychologists, Freud, Adler, Fromm, and Overstreet, offers some invaluable insights into the trouble with man's current relations with his brethren. Man's problems center on his lack of emotional maturity as exemplified by his glaringly evident predilection for emotional motivation and his avoidance of taking thought. Clarity of logic will be hard for men to achieve inasmuch as the prevailing institutions which influence us are so predominantly childish, self-seeking, and emotionally motivated. Perception of reality is obscured by one's

161

environmental conditioning to accept current popular belief as the highest good. The scientist and philosopher who critically seeks the truth knows all too well just how difficult it is to be honestly objective. And ultimate truth is all the harder to come by when "truths" of many colors and postures are imposed by national and religious groups.

Equally confused is the attempt to discern the hallmarks of emotional maturity. However, the Brotherhoods long ago recognized the highest values in human behavior and described these goals as the *virtues*. Now men outside the Brotherhoods have arrived at pretty nearly the same conclusions as a result of psychologists' recent efforts to ferret out the ideal characteristics exemplified by the completely mature man. We might expect a person of balanced emotional maturity to be considered the epitome of social acceptableness; but although he may be admired and respected, today's world remains aloof from him. Our society instead rewards the person who conforms to its current modes of behavior. Evidence of strength of character in a man marks him as a disquieting and irritating influence.

There has been so much popular misunderstanding about Freudian frustration that a trend has developed to indulge one's every whim lest suppressed desires lead to neurosis and psychosis. Freud never advocated hedonism as a psychological panacea, nor were his discussions related to conscious determination. Rather, he was concerned with conflicting drives on the subconscious level which the patient could not resolve consciously. Clinical experience among psychiatrists indicates that the current fads of self-indulgence, self-pampering, and self-dissipation have backfired into causing grave emotional illnesses. Life is by nature a continual series of frustrations and conflicts, and maturity is measured by one's ability to deal effectively with them as they arise. The development of character and wisdom is not enhanced by following emotional whims and group manias, nor can the adventure of life yield egoic advancement if the mind is clouded by alcohol or tranquilizers. Sanity grows upon the sharp point of contact with reality and keen alertness to its challenges; whereas avoidance of life and one's problems is delusory.

Individuality and moral integrity protect one from being stampeded with the multitude who surge blindly to the command of the impersonal though all-powerful fad-maker—*they*. On every side we are coerced to strive for social acceptance, and our children are made to sacrifice all too much on the altar of popularity. But to have achieved perfect adjustment to a society that is childish is to have regressed. Our social aspirations are presently geared to glamour, and young and old alike have become willing pawns to the "image makers." The manipulators of glamour and advertising have carefully designed these forces to make us discontent with what we are and possess so that we are moved to purchase and consume whatever goods are represented to advance one's status. As a result of this insidious conditioning, we worship celebrities who are renowned for little more than their well-knowness, glorify the executive and his attendant wealth symbols, and overemphasize sex for the sake of unattainable, out-of-this world romance. America has retreated from political, moral, and economic reality so that life as it really is and the rewards for living it on its own merits have been largely abandoned for a glamour world. This national hysteria is equivalent to Germany's erstwhile romantic hysteria of Supermanism. One's adeptness at playing at life according to the mode of American juvenilism is not to have reached the height of human aspiration.

We bend our energies to the pursuit of pleasure, property, and plaudits instead of to the pursuit of maturity. The former lead to folly and disillusionment whereas the latter brings contentment, confidence, and self-respect. Almost everyone thinks pleasure should bring happiness, and so it follows that pleasure has become highly desirable. By comparison, hazy misapprehensions about the rather rare state of psychological maturity restrict its popularity as a goal.

Admittedly, it is difficult to have aims beyond one's own horizons of understanding. Moreover, nobody can portray the advantages of maturity to a childish person because no amount of description can convey the feelings of an emotion to a man who has not already experienced that emotion. The short-range advantages of pleasure are apparent, but the long-range advantages of maturity are obscure and more difficult to realize; therefore, the majority of mankind has always accepted the quicker goal and has barely even considered the greater goal.

One of life's cruelest deceptions lurks in the shallow ambitions of man because his drives for pleasure, status, and ease are self-defeating. When they are satiated, the result is boredom, discontent, and dejection. The hoped-for happiness is a will-o'-the-wisp that leads the pursuer to seek newer pleasures after each in its turn proves as devoid of innate happiness as the last. Nonetheless, perennial happiness *is* possible to the possessor of psychological maturity. Not so much because of the maturity *per se* but because of the attitudes that made maturity possible. Selfless labor for others, love of humanity, love of nature, active furtherance of high principles, and communion with divinity are the demonstrated wellsprings of undiminishing human joy. Although myths must eventually bow to realities, man is more emotional than rational; so the shortsighted goals still prevail despite all the evidence against happiness being derived from pleasure. Literally thousands of novelists have hammered away at this human failing to correctly relate cause and effect, but their vivid accounts of men's futile and perverse struggles for happiness seem to have been largely discounted by their readers.

The philosophers of the 18th century were stirred to great hope for mankind when the concept of *reasonable man* was propounded. The application of scientific inquiry to the enigmatic phenomena of the physical world had brought about the dramatic emergence of order out of confusion and gave rise to the belief that the puzzle of human behavior would

164

likewise be reduced to rules of mathematical clarity and precision. Their unscientific idealistic hope has since proved wrong—at least by their standards. The ground rules for science were drawn up during the 18th century and were based on the lowest common denominator of materialism. Testimony offered via supramentality and transcendent sensory perception are not admissable evidence to clarify man's cause for existence, and so mankind still gropes in vain because of restrictions fixed by himself. Thus scientists are shackled by what traditionally cannot be done. The scientific mind when freed of these nonvalid restrictions and then tempered with parapsychology may yet lead man to new and greater philosophical tools. Americans scornfully point to the absurd limitations placed upon Russia's intellectuals and scientists by dialectical thinking, yet they fail to see the extent of the limitations in traditional Western thinking because they cannot view themselves as objectively.

As far as the 200-year-old dream of *reasonable man* is concerned, it has been attained only limitedly. I would venture to say that far less than one percent of the world's population approaches the mark. Upon these few persons rests the stability of society, and this is all the more remarkable since few hold high positions of temporal power. Perhaps no more than 10,000 men and women of the upper echelon of reasonable maturity keep the world on even keel. Most of these 10,000 are educators, philosophers, moralists, and humanitarians; and they, together with an unintended following of perhaps fewer than one million persons of goodwill and practical good sense, comprise an informal association for the preservation of mankind. All the rest of mankind are drifters who miserably fail to sustain civilization if the percentage of mature individuals within their society falls to a low percentage of the population.

Less than one person in 2,500 can be considered emotionally mature, and only this *aristocracy of excellence* is really ready for democracy. Even in the democratic nations of Western civilization the average citizen sells himself out at the polls. I believe Freud was right when he asserted that democracy will fail because of the emotional flaws in man. Abraham Lincoln prolonged democracy on earth by being dictatorial at the crucial moment, but the British and American traditions of democracy continue to be undermined by universal suffrage. Western civilization is producing some of the most brilliant men and women to be seen for many millennia, but decadent and irresponsible individuals are multiplying far more prolifically and consequently so is their voting power. The hard-won freedoms of the democratic governments are being thoughtlessly surrendered by "emotional peasants" in exchange for "security."

Despite these seeming drawbacks, the Brotherhoods insist that a democratic form of government is best for mankind; but they admit it can survive only within a citizenry almost wholly composed of emotionally mature individuals. The Brotherhoods hope to prove their contention by assembling the truly capable and mature persons of the world into a single group. If men of maturity and wisdom fail to unify themselves in this way soon, it could well mean the end of political and philosophic freedom anywhere on earth for all time to come. Future seismic activity and political disintegration should serve to alert the mature individuals of the world who have been tardy in perceiving the imperative importance of so uniting. Even now eminent social and economic analysts are warning us of many destructive trends that indicate a likely collapse of our present way of life in the very near future. I suppose I too could be classed as a prophet of doom, except that I foresee the wonderful nation-to-come being prepared under expert guidance. Whatever dire cataclysms man brings upon himself for the remainder of this century can be tolerably viewed by understanding that it is for the greatest good in the long run.

Our sick world is beset by problems that cannot be solved by the conventional power plays of times past, and I am doubtful that enough politicians throughout

166

the world will adopt attitudes likely to solve the dilemmas of our times. At any moment statesmen may thrust us into a crisis that can swiftly compound into a debacle of unimaginable destruction. That is a sickeningly pessimistic outcome to visualize, but it is realistically probable. I would like to console myself with a Pollyannish dream that mankind's sanity will prevail to save civilization, for then I could retreat into blissful disregard of the armament race. But reasonable men are unable to appreciate the motives of madness and therefore they have been repeatedly overwhelmed by lunatics. Madness will be at the helm during most of what remains of this century, and it is painfully obvious that we are particularly unable to control madness in other countries.

A sense of hopelessness is coming more into the open as we talk to the man on the street. He still has goals for himself, but his long-range dreams for mankind are very tentative. He advises others to grab what they can out of life while life still exists. He extends this prerogative to politicians and surrenders his concern for the outcome of the fiscal policies of his government. Poll takers have uncovered the average citizen's willingness to risk nuclear war, and this points up the way in which the emotional tautness of unremitting anxiety can fray the lines of rationality. There are still mature adults striving to avert conflict, but powerfully vocal groups in both the communist and capitalist camps are all for having it out and done with.

The relatively few years of peace since the end of World War II have brought a great increase in prosperity and technological improvement to both capitalist and communist countries. The world is big enough for both ideologies to work out their problems side-by-side; and if they would not harass each other, a century of mutual noninterference could see each type of government moderating toward the other in form. The communist despots are beginning to see that the profit incentive for individuals is essential for meaningful economic growth, whereas the democratic societies are slowly socializing. If the smaller nations emerging from feudalism were not constantly badgered by the Eastern and Western blocs, they would likely develop forms of government somewhere between the two extremes. The people of the backward lands are not so much concerned with political ideologies as with the quickest way to achieve the same advantages of materialism exemplified by the United States of America. They want food for their bellies, advanced medical care and public health, and they aim to have it by whatever means will procure it the quickest. Since there are glaring failures in both the Soviet and capitalist forms of so-called democracy, neutral nations should be left free to devise middle-of-the-road political systems. They might come up with some practical innovations in self-government which would be worthwhile for the larger nations to adopt. A number of small nations thus could serve as empirical proving grounds for democratic refinements. In any event, a hands-off policy by the large nations is essential and proper if for no other reason than that interference in the internal environment of a people is karmically disastrous and gallingly impertinent.

We needn't worry about my views becoming an eventuality because there is not enough trust between the big nations to give it a try. Trust is precluded by the communists' deceitful propaganda, world-wide subversive activities, and avowed determination to destroy the United States of America. The Western powers dare not relax a moment in the face of such a relentless diabolical enemy. And this enemy is all the more dangerous because dedicated communists exude the fervor of evangelism in an era when most of the suppressed peoples of the world are clamoring for a change in the status quo. The Western concept of democracy is the better way, but it is now the old way that in its time failed to improve conditions for the masses of most of the world. To this extent communism has acted as the conscience for the shortcomings of colonial capitalism. We would like to see the nations which are newly liberated from colonialism adopt democracy as their form of government, but their peoples are sorely unprepared for intelligent self-rule. The confused and unsophisticated people are an easy mark for communist propaganda which offers golden promises couched in terms of a glorious adventure in achieving a social paradise. By comparison, the Western bloc's lack of a purposeful, positive goal around which to rally its own people and inspire the world leaves us disunited and bereft of dynamism. Communism is winning the battle of idealogies for the same reason that Christianity triumphed in the Roman Empire—idealism and great expectations. The communists are remarkably successful in imposing their philosophy upon others; and when the Western nations finally find themselves backed against the wall in a losing battle of wits, they are likely to resort to force like any cornered creature.

There are many thoughtful persons who are sorely saddened by the prospect of seeing human beings reduced to a stone-age existence again. The likelihood of one's grandchild (should one survive a nuclear contest) being reduced to a stupid, snarling, short-lived, malnourished clod is crushing to one's sensitivities. One cries out in anguish at the thought of mankind's achievements going to naught. Yet no one is more sympathetic to this pathetic state of affairs than the Brotherhoods, for they have watched civilization obliterated many times over. In certain instances, men of quality who have cried out against the insanity of it all have been granted solace by the Brotherhoods who took them in and enlisted their talents.

If the average man of today were transported to a civilization like that of ancient Lemuria or the coming Kingdom of God, he would be a jarring misfit. His selfishness, his negative desires and ambitions, his arrogant intolerance, hatreds, cruelties, and overall lack of personal peace and happiness would make it impossible for him to get along in cooperative harmony. Since fear, anxiety, fretfulness, and irritability in one individual tends to spread to those who come in contact with him, it is obvious that he would only de-tract from the spiritual peace of a Lemurian environment. In spite of the luxuries, beauty, and material abundance that a society can create, its civilization is vapid, restless, and unfulfilling if the citizens are no better than the man of today. He has not learned self-discipline let alone self-government. He would require too many laws, and the Lemurian way of life relies on its code of ethics without external enforcement.

The very minimum of government is a Lemurian principle lest the government operate in the environment of the citizens against Cosmic Law. The purpose of Lemurian government is to provide public services inexpensively; and although there are government administrators in the Lemurian scheme of things, there will be no politicians. Democracy will be exercised by referendum and not by a republican legislature bending to pressure groups and lobbies seeking legislative favoritism; nor will organized political parties vie for control of government policies. The executive officers of the government will not be the leaders of the people but will be strictly public servants having supervision over only their respective departments and offices. The people will look to the great philosophers among them for inspiration and counsel, and then they will act for themselves.

Lemurian economic policies will seem quite different from the traditions we follow today. Machinery will put an end to hard labor, and automation will reduce the workweek to a few hours. This will result in low weekly wages, but the cost of such necessities as food and rental of housing will permit the wage earner a comfortable surplus for savings and luxuries. Automated farms will produce food very inexpensively, and radically simplified distributions methods without middleman markups will hold food prices close to the cost of rain, sun, fertilizer, and seed. Hard goods will be made to last for generations so that natural resources are conserved. Designs in the styling of appliances and autos will be such that they will look attractive indefinitely as well as outlast the

170

owner's lifetime. Production methods today are geared to produce the greatest profit for the stockholders; therefore, continual style changes accompanied by advertising that has conditioned us to scorn the old and worship the latest makes us discard what is still useful. Just in case some persons might prefer to get the maximum usage out of an item, defects are purposely engineered into it so it will need to be replaced early. The savings involved when one needs to buy only one automobile instead of ten during his life is considerable.

The Lemurian economic system is designed to fulfill *all* the mundane needs of *every* citizen. Industry will be due its profits but making profits will not be the attitude impelling manufacture; nor will people be driven to "keep up with the Joneses" or seek social status through high income or elite possessions. The long period of serviceability of housing, autos, furniture, and appliances will allow later generations to live their lives without having to replace goods passed to them by their forebears. A tremendous savings in resources and human labor will result when only food, fuel, clothing, and personal services need be purchased. Our present system is designed to keep us from catching up, and it has come to depend upon the exchange of goods and services at maximum production capacity in order to avoid collapse.

The great increase in the amount of leisure time to be enjoyed in the Lemurian system will be occupied by schooling. Education will be a lifetime avocation in the Kingdom of God, and citizens will be disposed to eagerly seek knowledge of all things. Every effort will be made to perfect education as well as the socio-economic environment so that the door to spiritual advancement for man will be opened to the Nth degree. The individual's attainment of Adeptship in the Brotherhoods will be the end goal of every activity in the Kingdom of God, and the beauty, bounty, and tranquillity of that nation will be regarded primarily as enhancement to egoic perfection.

The Kingdom of God has been planned by the

171

Brotherhoods for thousands of years, and it shall be successful from the inception of the community in Illinois. Perhaps at last we shall see to what heights man can rise in fulfilling his quest for civilization.

As you can ascertain, if the philosophy that I adhere to holds the correct beliefs about our future, the job ahead is going to be extremely difficult, though highly rewarding if successful. Perhaps the nicest thing about this philosophy lies in its perception of man's tremendous potential. Man has all the tools needed including the qualities of mind and resources provided with this planet. Perfection is possible but it is attained with effort, not doled out sporadically to this person or that.

Another reassuring attribute of this philosophy is that there is no such thing as "blame." If a person makes mistakes even of such magnitude as to be judged devastating by others, he alone is responsible and the karma inherent in Universal Law will exact appropriate retribution as a propitious learning device. But even the Masters have been along the path of the everyday physical grind and have erred; they have forgiven themselves and so must we.

Though society has bad points, we need not dwell on our blunders. We need to accept our responsibilities and work positively to improve rather than take precious time out from our task to point the finger of blame. It is especially deleterious to blame ourselves for that would diminish us. More mistakes will be made but they will be overcome. It would be a shame to have all progress blanked out with the death of planet earth after its life had exhibited such an impressive level of growth.

EPILOGUE

The work you have just read was not intended to constitute a final solution of any kind. Rather it has been my purpose to trigger thought patterns that will undoubtedly spark original ideas in the minds of many. This has been a cursory work, and naturally all the aspects of past, present, and future need considerable detailing, but that is a job for each of us in our daily lives.

I owe the debt of gratitude to far too many tremendous thinkers to try acknowledging them individually; however, I will offer a brief list of some suggested reading which may help many of you make up your own minds about the veracity of my work.

The Sun Rises by Dr. Stelle is no longer in general circulation; the best way to find a copy is to browse in old bookstores. *The Ultimate Frontier* can be obtained, in paperback for $2.00 (prepaid) from The Stelle Group, Cabery, Illinois 60619.

A series known as *Sourcebooks* compiled by William R. Corliss is excellent; his firm is located in Glen Arm, Maryland 21057. These books are primarily compilations of old scientific articles written on various topics that would constitute anomalies, scientifically speaking.

On cataclysmology you may find *Worlds in Collision* and *Earth in Upheaval* by Immanuel Velikovsky interesting; also the writings of Dr. Charles Hapgood who is superb in my opinion, and *Cataclysms in the Earth* by Hugh A. Brown.

Recommending books is never easy because there are so many good ones to read; and, of course, there are flaws and faults and errors and slanted opinions in nearly everything published, so each reader is forced to employ the virtue of discrimination.

On Atlantis and artifacts indicating its existence there are many books to peruse, including the now re-issued works of Ignacius Donnelly, the excellent *Atlantis Discovered* by Lewis Spence and, of course, the two mentioned in the text of this book by Robert Stacy-Judd and Anton Braghine.

The Great Pyramid—Man's Monument to Man is my own effort at unraveling the mysteries of that great structure, and Peter Tompkins' *Secrets of the Great Pyramid* is an excellent source of information. Most of the "pyramid power" books are interesting, and while they are not yet very scientific, the subject will one day be better understood.

On the workings of "evilists," you may find *The Screwtape Letters* by C.S. Lewis quite revealing.

Like everyone else, you take your chances when you start delving into metaphysics and the so-called field of occultism. However, I found the series of "Don Juan" teachings by Carlos Castaneda especially valuable when all four books are read in the order written.

In all of our endeavors to learn and to grow I believe one thing must remain steadfast—individual responsibility. There is no such thing as a singular way or an "only" way to attain our purpose. Though the universe is obviously from a single source, the fragmentations of that source are diverse and individualized and must work their way back on their own accord.

I hope I have conveyed that essential message more than any other with this work. We have free will, and if we strive we can also have greater knowledge and better understanding.